Center for the Study of National Reconnaissance Classics

A HISTORY OF THE HEXAGON PROGRAM

CENTER FOR THE STUDY OF
NATIONAL RECONNAISSANCE
CHANTILLY, VA

APRIL 2012

Foreword

This volume re-publishes, *A History of the Hexagon Program—The Perkin-Elmer Involvement* as part of the *Center for the Study of National Reconnaissance's (CSNR) Classics* series. The introductory information explains the corporate perspective and technical aspects of this history. Reading about the technical aspects of the program should help the reader understand how this program was able to make valuable contributions to U.S. national security. The history cites examples such as SALT verification, coverage of crisis areas like the Middle East, and terrain mapping for the Cruise Missile.

In 1983, the day that Perkin-Elmer shipped the twentieth Hexagon Sensor Subsystem to the West Coast, the then Director of the National Reconnaissance Office (NRO), Edward C. Aldridge (who also was Under Secretary of the Air Force), told a group of Perkin-Elmer employees, "The success of the Hexagon Program has established the standard for all future satellite reconnaissance programs to emulate and a goal for our present systems." This history will offer you insight into the technology that made that possible.

The *Center for the Study of National Reconnaissance Classics* is a series of occasional CSNR publications whose purpose is to inform our readers about classic issues from the past. The books and monographs in the series most typically are histories, but they also could address lessons-learned topics, the legacy recognition of people and programs, insights into historically significant artifacts, or tutorials on the discipline of national reconnaissance. We issue the publications in the series on both an *ad hoc* basis, or in connection with a significant event. We are issuing a Gambit-Hexagon collection of histories in response to Director of the NRO Bruce Carlson's decision in June 2011 to declassify the programs and his subsequent declassification announcement on 17 September 2011. The Historical Documentation and Research (HDR) Section of the CSNR selected five classic histories of the Gambit and Hexagon programs:

- *A History of Satellite Reconnaissance—The Perry Gambit & Hexagon Histories* (by R. L. Perry)
- *The Gambit Story* (by F. C. E. Oder, J. C. Fitzpatrick, & P. E. Worthman)
- *The Hexagon Story* (F. C. E. Oder, J. Fitzpatrick, & P. E. Worthman)
- *Hexagon Mapping Camera Program and Evolution* (M. Burnett)
- *A History of the Hexagon Program—The Perkin-Elmer Involvement* (by R. J. Chester)

On 21 January 2012, the CSNR published the first volume in the Gambit-Hexagon CSNR Classics series, *A History of Satellite Reconnaissance—The Perry Gambit & Hexagon Histories*. We did this in support of the ceremony that marked the NRO turning over a collection of Gambit and Hexagon artifacts to the National Museum of the United States Air Force (NMUSAF) and their exhibit opening of these artifacts to the public. The opening of this exhibit represented the largest collection of satellite reconnaissance artifacts ever assembled and put on public display. That exhibit can serve as a companion resource to those who read the histories in this CSNR Classics collection.

Each of these histories offers a different perspective on the programs; the Perry Gambit and Hexagon histories are from the viewpoint of a former Air Force historian at RAND writing in response to tasking from the then NRO Program A (Air Force program); the Oder, et. al. Gambit and Hexagon histories are from the viewpoint of authors with program experience working under the sponsorship of the Deputy Director of the NRO; the Burnett Hexagon mapping system history is from the viewpoint of the Hexagon program office working under the direction of two Air Force officers in the program and the NRO Program A Director; and the Chester Hexagon history is from the viewpoint of Perkin-Elmer, which was an associate contractor for the Hexagon program.

All of the authors researched and wrote their histories during what some observers might describe as the height of the Cold War, from 1964 to 1985. This influenced them to react to and focus heavily on the threat from the former Soviet Union and its allies. Also, all of the authors had at least some degree of first-hand knowledge about these programs, and in many cases, they had first-hand experience working in the programs. This gives you a window into what it was like to be a participant-observer in the development and operation of these film-return satellite photoreconnaissance systems during the Cold War.

Dr. James D. Outzen, the NRO Senior Historian and Chief of the CSNR's HDR section, is the editor for the Gambit-Hexagon CSNR Classics series. Dr. Outzen selected the five histories for this CSNR Classics series from the NRO Records Center and CIA archives that collectively best retell the impressive Cold War story about these programs. He has prepared a brief preface and introduction for each history to provide context and explain its significance.

When you read the histories you will note that some information is missing. Even though the Director of the NRO authorized the declassification of almost all the programmatic information about these programs, some information, because of its potential impact on other sources and methods, remains classified. Dr. Outzen usually let the redacted text stand on its own, but in some instances he has done some editing for readability. For some of the histories, Dr. Outzen has incorporated supplemental reference material into the publication.

Robert A. McDonald, Ph.D.
Director
Center for the Study of National Reconnaissance

Preface

Coinciding with the commemoration of the 50th Anniversary of the National Reconnaissance Office (NRO), the Director of the NRO, Mr. Bruce A. Carlson, publicly announced the declassification of the Gambit and Hexagon imagery satellite systems on 17 September 2011. This announcement constituted the NRO's single largest declassification effort in its history. The Gambit and Hexagon programs were active for nearly half of the organization's history by the time of the declassification announcement. Their history very much represents the NRO's history—one that is defined by supremely talented individuals seeking state of the art space technology to address difficult intelligence challenges.

The United States developed the Gambit and Hexagon programs to improve the nation's means for peering over the iron curtain that separated western democracies from east European and Asian communist countries. The inability to gain insight into vast "denied areas" required exceptional systems to understand threats posed by US adversaries. Corona was the first imagery satellite system to help see into those areas. It could cover large areas and allow the United States and trusted allies to identify targets of concern. Gambit would join Corona in 1963 by providing significantly improved resolution for understanding details of those targets. Corona provided search capability and Gambit provided surveillance capability, or the ability to monitor the finer details of the targets.

For many technologies that prove to be successful, success breeds a demand for more success. Once consumers of intelligence—analysts and policymakers alike—were exposed to Corona and Gambit imagery, they demanded more and better imagery. Consequently, the Air Force, who operated the Gambit system under the auspices of the NRO, entertained proposals for an improved Gambit system shortly after initial Gambit operations commenced. They received a proposal from Gambit's optical system developer, Eastman Kodak, for three additional generations of the Gambit system. Ultimately the Air Force settled on only developing the proposed third generation because the proposed second generation offered minimal incremental improvement and the fourth generation appeared technologically unachievable at the time. The third generation became known as Gambit-3 or Gambit-cubed while it was under development. Once it replaced the first generation, it simply became Gambit. The new Gambit system, with its KH-8 camera system, provided the United States outstanding imagery resolution and capability for verifying strategic arms agreements with the Soviet Union.

Corona was expected to serve the nation for approximately two years before being replaced by more sophisticated systems under development in the Air Force's Samos program. It turned out that Corona served the nation for 12 years before being replaced by Hexagon. Hexagon began as a Central Intelligence Agency (CIA) program with the first concepts proposed in 1964. The CIA's primary goal was to develop an imagery system with Corona-like ability to image wide swaths of the earth, but with resolution equivalent to Gambit. Such a system would afford the United States even greater advantages monitoring the arms race that had developed with the nation's adversaries. The system that became Hexagon faced three major challenges. The first was development of the technology, which was eventually overcome by the Itek and Perkin-Elmer Corporations. The second was bureaucratic, deciding how the CIA and Air Force would cooperate in building such a system because they each had strengths and weaknesses in the development of national reconnaissance systems. The third challenge was to secure the resources that were required to build the most complicated and largest reconnaissance satellites at the time. By 1971, the NRO overcame the challenges to successfully launch the Hexagon satellite and fulfill, or even exceed, expectations for unparalleled insight into capabilities of US adversaries.

At the time of the Gambit and Hexagon declassification announcement, the NRO released a number of redacted Gambit and Hexagon documents and histories on its public website. One of the histories is contained in this volume.

A History of the Hexagon Program was written in 1985 by Richard J. Chester of the Perkin-Elmer Corporation. Perkin-Elmer took over development of the original camera system that would evolve into the primary camera for the Hexagon program. The panoramic camera system, designated KH-9, would image vast areas of the earth and prove an essential collection asset for verifying the Soviet Union's compliance with arms control treaties as well as address other intelligence questions.

A History of the Hexagon Program is very rich in technical detail. Chester carefully documents Perkin-Elmer's efforts to develop the KH-9 camera system and integration with the other components that would become the Hexagon system. Chester also documents the contributions of many individuals who were responsible for the program's success. Chester includes a trove of drawings, photographs, and other visual content that enhance the narrative descriptions of Perkin-Elmer's contributions to the Hexagon program.

A History of the Hexagon Program is a unique program history in comparison to others written about national reconnaissance programs. Chester documents the corporate perspective of satellite program development. Compared to histories written by government contractors, Chester pays less attention to the politics surrounding competing broad area search satellite proposals. He also pays less attention to the intelligence needs for the satellite. Despite the diminished attention to these issues, Chester tells the technical development story in an unparalleled fashion. Consequently, we read in the pages of this volume one of the most complete retellings of the efforts to develop and integrate the components that comprise a reconnaissance satellite system.

A History of the Hexagon Program joins five other volumes of Gambit and Hexagon histories that the Center for the Study of National Reconnaissance is reprinting in conjunction with the program declassifications. Those other volumes include *The Gambit Story* and *The Hexagon Story* both written by Frederic Oder, James Fitzpatrick, and Paul Worthman, Robert Perry's histories of Gambit and Hexagon, a history of the Hexagon mapping camera, and a compendium of key Gambit and Hexagon program documents. In total, this collection of Gambit and Hexagon publications provides the public with broad insight into previously classified programs. The volumes complement each other in providing details not found exclusively in any single program history volume.

At the time of this writing, KH-9 panoramic camera system imagery has not been declassified. I have included in a separate section of this publication a small number of KH-9 images that were released in conjunction with the Hexagon declassification.

I have chosen not to reprint pages that were redacted in their entirety in *A History of the Hexagon Program*. Those pages are: 176 –180. The unedited redacted *A History of the Hexagon Program* can be found in the declassified records section of NRO.gov for those interested in reviewing a document with the completely redacted pages.

The Gambit and Hexagon systems became reliable means for addressing difficult intelligence challenges once they became operational. The Hexagon system, in particular, provided broad area imagery that was essential for understanding the strategic capabilities and arms control compliance of the Soviet Union and other Cold War adversaries. These national reconnaissance systems dutifully provided the nation reliable vigilance from above until the next generation of imagery satellites advanced US intelligence collection capabilities.

James D. Outzen, Ph.D.

Chief, Historical Documentation and Research
The Center for the Study of National Reconnaissance

Center for the Study of National Reconnaissance

The Center for the Study of National Reconnaissance (CSNR) is an independent National Reconnaissance Office (NRO) research body reporting to the NRO Deputy Director, Business Plans and Operations. Its primary objective is to ensure that the NRO leadership has the analytic framework and historical context to make effective policy and programmatic decisions. The CSNR accomplishes its mission by promoting the study, dialogue, and understanding of the discipline, practice, and history of national reconnaissance. The CSNR studies the past, analyzes the present, and searches for lessons-learned.

NRO APPROVED FOR
RELEASE 17 September 2011

Hx ~~TOP SECRET~~

THIS DOCUMENT CONSISTS OF <u>226</u> PAGES

A History Of The HEXAGON PROGRAM

**WARNING– THIS DOCUMENT SHALL
NOT BE USED AS A SOURCE FOR
DERIVATIVE CLASSIFICATION**

NATIONAL SECURITY INFORMATION
UNAUTHORIZED DISCLOSURE
SUBJECT TO CRIMINAL SANCTIONS

DERIVATIVE CL BY: BYE-1
DECLASSIFY ON: OADR
DERIVED FROM: BYE-1

COPY <u>2</u> OF <u>25</u> COPIES

BIF 007-0253-85
HANDLE VIA BYEMAN
CONTROL SYSTEM ONLY

Hx ~~TOP SECRET~~

THIS PAGE INTENTIONALLY LEFT BLANK

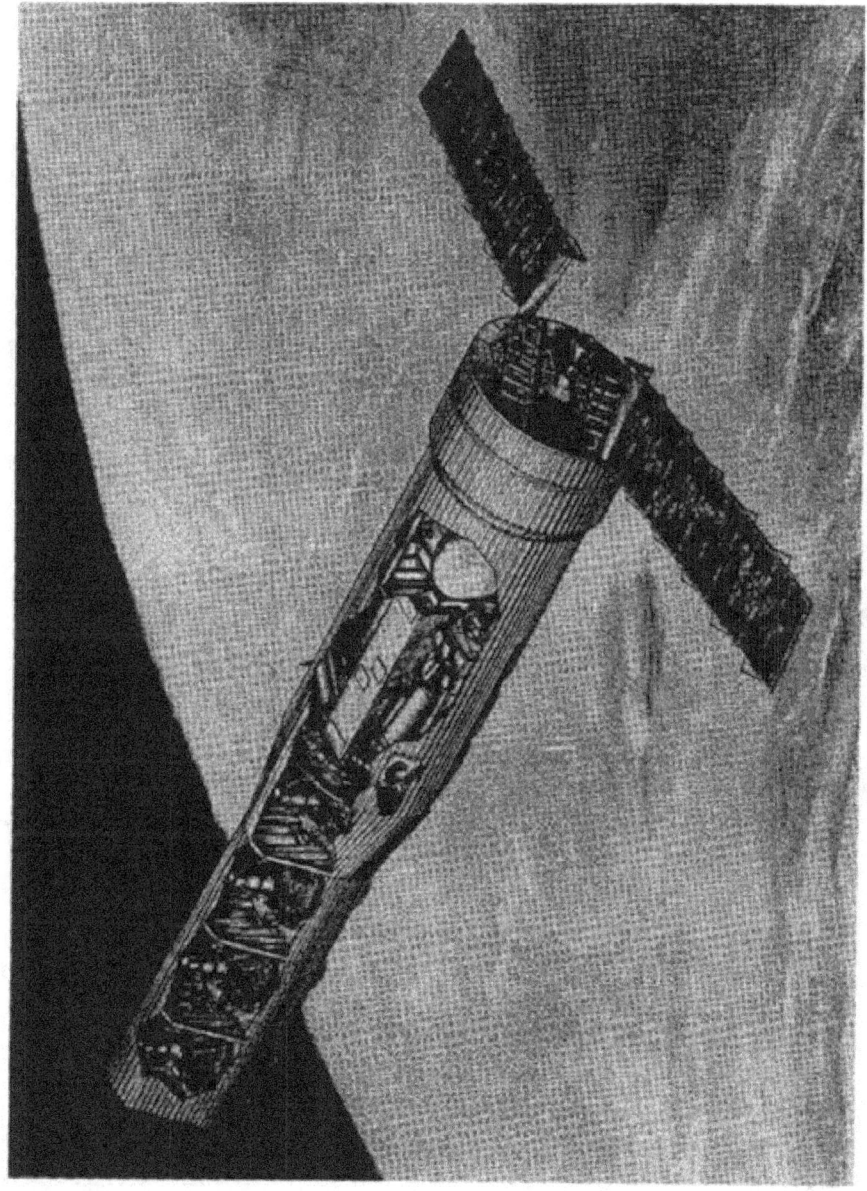

Hexagon Satellite Vehicle (SV-17 Configuration)

Hx TOP SECRET

*To
Richard S. Perkin
1906-1969
and the men and women who built
the Hexagon Satellite Vehicle*

Prepared by Richard J. Chester

BIF 007-0253-85
HANDLE VIA BYEMAN
CONTROL SYSTEM ONLY

Hx TOP SECRET

Hx ~~TOP SECRET~~

A HISTORY OF THE HEXAGON PROGRAM

Contents		Page
	FOREWORD	vii
	INTRODUCTION	ix
1	PROGRAM OVERVIEW	
	Early Background	1
	Organizational Period	12
	Early Technical Development	14
	Award of Contract	31
	Cover and Security Considerations	40
	Building Program	45
	Sensor Subsystem Description	49
	First Flight of the Big Bird	57
	Sensor Subsystem Improvements	60
2	CUSTOMER RELATIONSHIPS AND INTERFACES	
	Program Management	65
	Customer Changeover	66
	Program Security	67
3	TECHNICAL DESIGN, MANUFACTURE AND TEST	
	Evolution of the Sensor Subsystem Design	73
	Optical Bar Assembly	74
	Camera Support Frame Assembly	76
	Film Drive Assembly	80
	Platen Assembly	84
	Supply Assembly	88
	Looper Assembly	91
	Film Path Assemblies	94
	Take-Up Assembly	108
	System Electronics	110
	Systems Engineering	112
	System Reliability	128
	Manufacturing and Test	130
	Optical Fabrication	142
4	RELATIONSHIPS AND INTERFACES WITH ASSOCIATE CONTRACTORS AND SUBCONTRACTORS	
	Associate Contractors and Responsibilities	145
	Selection of Subcontractors	145
	Development of Interfaces	147

BIF 007-0253-85
HANDLE VIA BYEMAN
CONTROL SYSTEM ONLY

A HISTORY OF THE HEXAGON PROGRAM (Continued)

Contents		Page
5	SYSTEM INTEGRATION, LAUNCH, ORBITAL OPERATIONS, AND RECOVERY	
	Development of the West Coast Field Office	151
	Final Assembly and Testing of Flight Model 1 (SV-1)	153
	Mission Activities	157
	███████████████████	159
6	EPILOGUE	169
7	HEXAGON IMAGERY	175
APPENDIX A	Flight Performance Record	182
APPENDIX B	Mission Durations	183
APPENDIX C	Mission Operations	184
APPENDIX D	Imaged Photographic Frames	185
APPENDIX E	System Performance Trends	186
APPENDIX F	Photographic Coverage	187
APPENDIX G	Listing of Post Flight Reports	188
APPENDIX H	Organization Chronology	189
APPENDIX I	Description of Program Documentation	192
APPENDIX J	Source Notes	193

FOREWORD

At precisely 10:41 A.M., Pacific Standard Time, on June 15, 1971, the first of America's fourth generation photographic reconnaissance satellites (Hexagon program) lifted off a pad at the Vandenberg Air Force Base in southern California.[1] This event was the result of 57 months of intensive effort to design, manufacture, and assemble a revolutionary new intelligence collection system. Thousands of scientists, engineers, technicians, and administrators in various government and industrial facilities throughout the United States were involved in making it happen.

On July 12, 1983, the day the 20th Hexagon Sensor Subsystem was shipped from Perkin-Elmer to the West Coast, The Honorable Edward C. Aldridge, Under Secretary of the Air Force, Director National Reconnaissance Office, addressed a large group of Perkin-Elmer people who had participated in the design and manufacture of the Sensor Subsystem. A part of his speech follows.

"In June of 1971, the first of a new breed of satellite reconnaissance systems was launched from Vandenberg Air Force Base and ushered in a new age in terms of satellite photography. The program success has continued almost unabated and permits us to talk in glowing terms of performance characteristics such as: up to seven months on orbit; over 300,000 feet of film (60 miles); total mission area coverage nearly equal to the landmass of the earth.

We didn't get to this point by accident; it was achieved through perseverance, technical competence and a lot of hard work. Many of you in the audience, I am sure, can remember the bustle of the initial years. In comparison, today must seem very serene. That initial launch in 1971 was not really the beginning but rather the culmination of a dream. This dream started in 1965 when your company became involved in a competition to build a revolutionary spacecraft capable of handling reconnaissance requirements fostered by the emergence of Russia and China as superpowers. In 1966, Perkin-Elmer was chosen to participate as an associate contractor in what has become one of our most important intelligence programs. It is hard to believe that seventeen years have passed since that date. However, the validity of that initial decision has been confirmed and reconfirmed with the passage of time. The success of the Hexagon Program has established the standard for all future satellite reconnaissance programs to emulate and a goal for our present systems.

This success can be measured in many ways. For example, since the first launch in 1971 67 out of 68 buckets have been recovered (see Editor's comment); imaging lifetime has increased over 800 percent; film capacity has increased 50 percent; mission area coverage has increased 100 percent. Even with these impressive facts, it is impossible to measure Hexagon's considerable contribution to our national defense. This importance can be sensed only if we cite examples such as SALT verification, coverage of crisis areas like the Middle East, and terrain mapping for the Cruise Missile."[2]

The Honorable Edward C. Aldridge, Under Secretary of the Air Force

Hx TOP SECRET

This history tells the story of how Perkin-Elmer was selected to participate in the Hexagon program. It relates the circumstances that led to Perkin-Elmer's involvement in space reconnaissance and discusses the events and milestones that occurred on the Hexagon program to the end of 1983. It reveals Perkin-Elmer's relationships with government program managers, the associate contractors working on the project, and the major subcontractors who helped to build the equipment. It reviews the evolution of the design of the camera and the film transport system and the technical problems that were faced and solved.

You will be taken through a tour of the Perkin-Elmer facilities both in Connecticut and California and learn how a large space reconnaissance camera is assembled, tested, and integrated with the launch vehicle. And finally you will read of the excitement of the first flight and look at the sharp details that can be seen on photographs taken from an altitude of more than 100 miles. It is the story of the Hexagon Program from Perkin-Elmer's viewpoint.

Although the History of the Hexagon program will be read initially by a select few who can benefit by the information it contains, and perhaps apply this knowledge to existing and future programs, these pages will also preserve the story of the Hexagon program for the day that it can be revealed to the American people. Only then will they understand and appreciate the total dependence of our defense system -- and their freedom -- on the success of our space reconnaissance programs.

INTRODUCTION

The history of space reconnaissance may be thought of as three separate histories: the evolution of spacecraft, the development of reconnaissance equipment, and the transformation of the intelligence community into organizations largely dependent on automated orbiting electronic and optical sensors commanded and controlled by high-speed computers.

Compared to European and Asian nations, America is a newcomer to centralized intelligence. It was during World War II that we learned our future survival depends on our ability to anticipate the moves of aggressive and expansionst countries.[1]

On January 22, 1946, President Truman established the Central Intelligence Group to act as a central repository for intelligence information and to analyze and fit all the pieces together into a coherent picture. But the Central Intelligence Group merely added one more report to those already flowing over his desk. The real change came on July 26, 1947, when the Central Intelligence Agency was formed by the National Security Act. But this time it was different. The agency had its own budget and was responsible to a new National Security Council.

In addition to unifying the armed services under a new Department of Defense, the National Security Act of 1947 established the United States Air Force, and in 1949, under a separate amendment of the act, the Air Force was made a military division within the Department of Defense with equal status to the Army and Navy.

Prior to and during World War II, the U.S. Air Force was a branch of the Army and consequently suffered inadequate support in funds, personnel, and in organizational status. It was not until November, 1940, that an Intelligence Division was created at the Chief of Air Corps staff level. This division became A-2 (Assistant Chief of Staff) within the newly organized Army Air Forces in 1941 with a responsiblity for both assessment and dissemination.[2]

Today, the National Reconnaissance Office (NRO) operates satellite programs for the intelligence community. The Air Force, with its large intelligence organization, supports the NRO, and has a special responsibility to guard against surprise.

Space reconnaissance had its genesis in a 324-page report published on May 2, 1946, by a group called Air Force Project Rand at the Douglas Aircraft Company in California, later to become the Rand Corporation. The report concluded that technology had advanced to a point where it was feasible to undertake the design of a satellite vehicle.[4] However several events had to occur before this work could begin; the invention of the transistor, the development of powerful rocket boosters, and the availability of high-speed computers. In June, 1956, the last piece fell into place. Rand scientists completed a report proposing a bold new idea for physically returning photographic film from orbit -- the reentry vehicle.[5]

By late 1958, plans were made to launch experimental recoverable satellites by the Thor/Agena space booster, given the family name Discoverer. At the same time, work began on the construction of launch and instrumentation facilities at the Cooke Air Force Base, California, subsequently renamed Vandenberg Air Force Base, and most recently given the title of Western Test Range. On February 28, 1959, the first Discoverer was launched into a successful orbit. Unfortunately, the satellite began to tumble and was lost, but it was a promising beginning.[6]

Most of the aerial photography in the early 1920's was accomplished by World War I cameras. It wasn't until 1921 that the K1 focal plane shutter, 9-inch film aerial camera was available. In 1958, the first photographic reconnaissance system to be designed and built under the weapon system concept was developed for the supersonic RB-58 Hustler aircraft; the KA25, KA26, and the KA27 cameras.[7]

Perkin-Elmer designed the Transverse Panoramic Camera, Type E-1 in 1952. It was the first camera of the panoramic type employing a rotating prism for scanning. Panoramic cameras offer a saving in weight over a multi-camera set-up for wide area coverage.

The U-2 Reconnaissance aircraft successfully utilized transverse-scanning panoramic cameras made by Perkin-Elmer during the 1950-1960 period; similarly, such designs were incorporated in cameras Perkin-Elmer built for the U-2's successor, the A11 aircraft. The vulnerability of aircraft, even high-altitude aircraft, to attack and the (mostly psychological) impact of intrusion into denied "air space" added urgency to the deployment of "untouchable" satellite-borne cameras having the same wide-area coverage, transverse-scanning designs.

Richard C. Babish, Project Engineer of the Transverse Panoramic Camera, Type E-1 in 1952.

By late 1960, the elements of space reconnaissance began to merge. The capsule of Discoverer 13 was the first man-made object recovered from space (see footnote*). Navy frogmen lifted the precious cargo from the Pacific Ocean and put it safely aboard the USS Haiti Victory. A week later, the Discoverer 14 capsule was the first to be recovered in midair as it descended by parachute over the Pacific Ocean.

In late 1965, the stage was being set for the final study of a new generation photographic satellite. It would be required to provide the resolution of earlier close-look satellites while simultaneously providing the broad area coverage capability of previous search/surveillance systems. On July 21, 1966 proposals for the Hexagon sensor were submitted to the government by both Itek and the Perkin-Elmer Corporation. At 1700 on October 10, Mr. Robert Sorensen, then Senior Vice President, Optical Group, received an important phone call from Mr. John J. Crowley, Director of Special Projects, CIA, -- Perkin-Elmer's proposal was accepted by the government.[8] This is a story of the events that followed.

*Of general interest is a book written by Alistair MacLean, "Ice Station Zebra", which uses space reconnaissance as its basic plot. Perkin-Elmer is mentioned (page 252) and reference is made to "another American firm" (apparently Itek).

1 PROGRAM OVERVIEW

EARLY BACKGROUND

It was early summer, 1964. Perkin-Elmer was in its 25th year, and in addition to its expanding commercial instrument business, it had successfully participated in numerous government programs, both classified and unclassified.

A young engineer, who managed the Engineering Department on a covert project at an outlying plant in Norwalk, Connecticut, was busily engaged in preparing a report. Things were going well on the program and he was formulating plans for final equipment tests. At that moment, the phone rang and the receptionist informed him that he had two visitors.

After hearing their names, Milton Rosenau, decided to greet them in the lobby instead of waiting for them to be escorted to his office. The two visitors were customer representatives on his program, Leslie C. Dirks and Jack Maxey. These three, all young and creative, had developed a special rapport during the life of their program. Milt Rosenau was a Cornell Graduate with a B.S. in Engineering and Physics. Les Dirks and Jack Maxey were Systems Analysis Staff Engineers in the Central Intelligence Agency (CIA) Directorate for Science and Technology.

Since this meeting had not been planned, Milt wondered about the purpose of their visit. After the usual amenities, Jack Maxey got into the real reason for their visit. A new reconnaissance program was underway and their agency felt that it might be to the mutual benefit of Perkin-Elmer and the Agency if Perkin-Elmer became involved.

Since the late 1940's, Perkin-Elmer designed many electro-optical instruments for both fighter and bomber aircraft. It was now time, said Jack Maxey, that the company consider getting into space programs. He hastily added that the current program had already been awarded to a Perkin-Elmer competitor and that Perkin-Elmer's role in the program would be to generate ideas that might be used to spur their vendor's thinking.

Milt's initial thought was that their approach to getting Perkin-Elmer involved was unorthodox but he realized that the highly secretive nature of the space program warranted special treatment. After their meeting was finished, Milt stopped off at his supervisor's office. Richard C. Babish, Director of Special Projects and one of Perkin-Elmer's most prolific engineers, listened to Milt Rosenau's story. Milt repeated the

The Perkin-Elmer building in which the initial contact was made involving Perkin-Elmer in space reconnaissance. This building also served as a base of operations for several aircraft reconnaissance programs.

customer's proposition and felt that although the program was already contracted, it was an opportunity for Perkin-Elmer to get involved with a major space program.[1]

Les Dirks and Jack Maxey did not reveal the exact nature of the program or their vendor. Perkin-Elmer would be given sufficient information to do their job - and no more. After getting the full details of the meeting, Dick Babish arranged a meeting that day with Dr. Kenneth Macleish, Vice President and Director of Engineering of the Electro-Optical Division. The Special Projects Department was part of that division. Dick repeated the customer's proposal. Dr. Macleish agreed that it was a wonderful opportunity and set up a meeting with the Vice President and General Manager of the Electro-Optical Division, Robert H. Sorensen.

This was the beginning of a chain of events which can best be revealed by excerpting from a memo written by Dr. Kenneth G. Macleish to the President and Chairman of the Board of Perkin-Elmer, Richard S. Perkin. It is dated September 19, 1964.

Subject: Chronology

The other evening you suggested putting down a history of the situation we were discussing. The following brief chronology covers the period from June 18, 1964 to date.

June 18 — Rod Scott and I went to Washington to talk with Jack Maxey and Les Dirks (of Bud Wheelon's staff) about a potential contract, at their request. Jack and Les outlined the broad requirements for a certain system and requested a proposal for a study program of about three months duration. The objective was to arrive at not more than two alternate configurations within 30 days and then to proceed with preliminary designs.

June 22 — Jack and Les visited us in the evening for further discussion of the proposed study.

June 28, week of (approximately) — A study contract was negotiated and work was begun. The final report was due September 28. Earle Brown was appointed project manager, to be guided by a committee consisting of Dick Babish, Rod Scott, Milt Rosenau, Bob Hufnagel, and myself.

July 6 (approximately) — A verbal progress report was given to Jack and Les in Washington.

July 27, evening — Verbal progress report to Jack and Les, in Wilton.

Aug. 14 — Frank Gorman, from Bob Greer's organization (Air Force) telephoned me to ask if we would care to propose on a study program for them. The requirements sounded like what we were already doing for Maxey but it was hard to tell for sure over the phone. I agreed to visit Frank the following week for further information.

Aug. 17 — Bob Lewis, Chester Nimitz, Bob Sorensen and I discussed the potential dilemma of being asked to do the same thing for two different agencies under security restrictions prohibiting us from informing either of the other's activity. It was obvious that we could not legally or ethically accept payment twice for the same work. On the other hand a gross undercharge to one or both customers would likely have to be explained, and such explanation might violate the security restrictions placed upon us. It was agreed that I should visit Frank Gorman as planned and find out more about the proposed new study.

Aug. 18 — Further discussions along this line between Bob Sorensen, Rod Scott, and myself. We decided that Rod should telephone Edwin Land (Chairman of the President's Science Advisory Board subcommittee on reconnaissance programs), tell him the story, and see if Land cared to advise us. Rod did so and reported that Land suggested we tell our first customer about the proposed new study. We decided to defer further action until after the visit to Gorman. I asked Rod to come with me on the Gorman visit. Since Milt Rosenau was going to be in that area, I arranged for him to come also.

Aug. 21 — The three of us were briefed by Greer's people and their technical advisers and asked to submit a proposal for a parameter study and preliminary design to be completed by February, 1965. It was expected that a go-ahead might be possible during the first week in September. In a conference after the briefing, Rod, Milt, and I unanimously agreed that the two studies were essentially independent and that no potential conflict or duplication existed. Our conclusion was based on the facts that (1) there was to be little or no overlap in time between the two, and (2) the technical requirements differed in many important respects which would entail much additional work on the second study over and above anything accomplished in the first one. We therefore decided to submit the requested new proposal without informing either customer of the other's work.

Aug. 28 - Sept. 10 — We submitted a proposal to the second customer and negotiated a contract.

Sept. 14 — We were visted in Wilton by Jack Maxey, Les Dirks, John McMahon, and John Crowley for a briefing on the first study. Jack said we had not proceeded as far as he had hoped toward deciding on a single configuration (we were still studying three) and asked that by September 28 we furnish him a proposal for three month's additional work. After much discussion he suggested that the follow-on study should include preliminary design of all three configurations plus some component development and experimentation.

Sept. 15 — ▉▉▉▉▉▉ of Bob Greer's organization telephoned me to communicate a change in one of the basic parameters in their specification. The parameter change brought the scope of the second customer's study very close to that of the first customer.

Sept. 16 and 17 — Various discussions were held among us concerning the implications of these recent events. It was clear that the changes proposed by both customers would bring about an overlap and duplication that had not existed previously. We therefore decided to tell Dr. McMillan (Under Secretary of the Air Force) or Jack Maxey or both what the situation was and ask for relief of some sort. Rod believed that by this time both Jack and Dr. McMillan were aware of both activities so that the danger of a security violation would have disappeared.

Sept. 18 (morning) — Bob Sorensen and I met with Maxey, Dirks, and McMahon in Washington. We told Jack we were trying to get an appointment with Dr. McMillan that day, and we told about the proposed study for Greer and our resulting problem. Jack revealed a need on his part for a considerably expanded and accelerated design effort and invited us to convince him, during the next few days, that we had the manpower to staff such an effort under an extension of our existing contract. He said that Dr. McMillan's office would be and had been receiving technical reports and regular briefings on his (Maxey's) activites.

Sept. 18 (afternoon) — Bob and I talked with Dr. McMillan. He said that McCone (Director of the CIA) had complained to him about engaging us to do the same work we were already doing for McCone. He asked if we could write him a letter stating that in fact there was no duplication. We said that up to a few days previously we could have written such a letter but that the rules had changed and there now appeared to be potential duplication. I briefly summarized the whole sequence of events described in this memo. Dr. McMillan said he had been aware of the other program, but had not known that we were involved in it. He seemed to understand our situation clearly and with sympathy and said he would call Bob Greer forthwith to get us off the hook. We said that we would like to continue working with Greer in any practicable way and we suggested that the scope of our proposed Greer contract might be modified to be a study of the specific application to his system of any configuration developed by us under the other contract. Dr. McMillan agreed that this might be possible and suggested that we communicate directly with Greer on the subject after he had made his own phone call.

Bob Sorensen and I are setting up a planning session for Monday morning, September 21, after which I plan to telephone Bob Greer. Maxey and Dirks will be available on September 23 and 24 to receive our proposal.

One who was not intimately involved in the situation spelled out in this memo will have difficulty "reading between the lines." Since the development of the U-2 reconnaissance aircraft and the camera equipment that it carried, there has been an ongoing institutional conflict between the CIA and the United States Air Force (USAF).

In 1955, soon after the first flight of the U-2, the operation of the reconnaissance program was assigned to the CIA. Since the CIA had no facilities, the USAF was assigned to take care of operations and logistics.[2]

This partnership carried over into the space reconnaissance programs when that same year the USAF, under the sponsorship of the CIA, issued a formal operational requirement for a strategic satellite program, code number WS-117L, the actual beginning of America's satellite reconnaissance programs.[3]

Before the story of Perkin-Elmer's involvement in this program can be continued, one additional aspect of this situation must be clarified.

The fourth participant in the early history of the Hexagon program is an optical company which was already involved in the design and manufacture of space reconnaissance cameras in the 1960's, the Itek Corporation. In the spring of 1964, Itek was approached by the CIA to develop an information-gathering system that could perform the operations of both the area-surveillance and the close-look satellites already in operation, with improved resolution and a capability for longer missions.[4] There was one significant problem that plagued photo satellites already in operation and that was thermal control. This was to become an important factor which would ultimately lead to the selection of a particular camera configuration.[5]

Itek completed a study (Itek project 9096) and concluded that an optical system working in the visible region of the electromagnetic spectrum was the best choice. Having made the basic selection, Itek submitted a proposal to the CIA (Itek proposal No. 3233) completed sometime in May 1964. However, before funding a full-scale development program, the CIA asked Itek to establish the feasibility of the critical aspects of their proposed system. This was accomplished in an Itek study (Itek project 9204) completed in 26 February 1965.[6]

Itek was also working with the USAF in their version of the fourth generation reconnaissance system. To understand the relationship between Itek and the USAF, it is helpful to have some knowledge of the formation of the Itek Corporation.

In the early days of World War II, the Air Corps took over one of Harvard's large library buildings, and formed an optical laboratory, manned by some of the finest optical designers and technicians in the country. Within a short time, lenses of every description were being developed and produced for the overseas forces. After the war, the laboratory was moved to the Boston University. Unfortunately, in 1957, the USAF reduced its funds for the support of the Boston Laboratory in favor of added support of radar and infrared developments.[7]

The laboratory was taken over by a team of top scientists in the Boston area and became the nucleus of the new Itek Corporation. So there existed a strong bond between the USAF and Itek. One additional factor strengthened this bond.[8] Brigadier General George W. Goddard, USAF retired, became a special assistant to the president of the Itek Corporation in 1964. Goddard, through World Wars I and II and the Korean conflict, was instrumental in developing many of the techniques and equipment fundamental to aerial reconnaissance.[9]

We can now resume our story of the events that occurred when the CIA contacted Perkin-Elmer in June 1964. As a result of the meeting on 18 June 1964 among Drs. Roderic Scott, Kenneth Macleish and the CIA, a study contract was awarded to Perkin-Elmer, via a message on 23 June 1964, to begin work under Task 4 to contract CH-900. The contract authorizing the work was received on 30 June 1964.[10]

As mentioned in Dr. Macleish's memo to Mr. Richard Perkin, Earle Brown was selected as Program Manager of the new study.[11] A project team was formed and named the "Ad Hoc" project. (The CIA code name for this project was "Fulcrum.") It consisted of 12 of Perkin-Elmer's most competent camera designers. In addition to Earle Brown and Milton Rosenau, the group included William A. Welch, John L. Rawlings, Roy Stoll, Robert M. Landsman, Michael H. Krim, Graham F. Wallace, Walter Augustyn, C. Donald Cowles, Karl W. Hering, and Walter McCammond.[12]

The program was planned in three steps: a parametric study of a photographic system to accomplish the desired objectives; investigation of alternate configurations within the constraints of size, weight, and other factors imposed by the vehicle and the mission; and a determination of the significant problem areas involved in the final choice. The original intent of the study was to complete all three steps and submit a final report by 28 September 1964. However, work did not proceed at the anticipated pace and on that date, the program was nearly through Step 2, with preliminary work accomplished on Step 3.[13]

A preliminary survey of possible configurations by the Ad Hoc project team resulted in the selection of three panoramic configurations, code-named Matchbox, Scarecrow, and Ferris Wheel. The initial effort was directed toward identification of representative designs capable of high resolution in both refractive and catadioptric types. Of all the systems considered, the folded Maksutov, and an unfolded variation of it were found to be most suitable. At that point in time, however, no sound technical basis had been established for the choice of one of these alternate systems.[14]

During a visit to Perkin-Elmer on 14 September 1964 by Maxey, Dirks, McMahon and Crowley, Maxey professed disappointment at Perkin-Elmer's lack of progress toward a single system and asked Perkin-Elmer to furnish the agency with a proposal for three-month's additional work.[15]

Meanwhile, the Itek Company had embarked on 1 September 1964 on a study (Itek Project 9204) which was to be completed 26 February 1965.[16] On 18 September 1964, the day that Mr. Sorensen and Dr. Macleish informed the CIA of Perkin-Elmer's proposal on the USAF project, Maxey revealed a need for a considerably expanded effort and accelerated design effort with a much larger engineering staff.[17]

The extent of the CIA program and the progress being made in other areas (i.e. spacecraft and reentry vehicle) was revealed by a memo received by Earle Brown from Leslie Dirks (dated 8 July 1964). It stated that on 1 September 1964, Phase I of a program to demonstrate technical and mechanical feasibility of the overall program had been initiated and was scheduled to be completed on 31 January 1965. The selected vendor would begin

The initial studies on the AD HOC Project were conducted in Perkin-Elmer's Research and Engineering Center, 50 Danbury Road, Wilton, Connecticut in June 1964.

development and production of flight units on 1 March 1965. The first launch was to occur in November 1966, with three flight tests (one per month) and subsequent operational flights (one per month).[18]

The proposal exhibit attached to the Dirk's memo also contained the first reference to a System Engineering, Assembly and Check-out (SEAC) contractor. This group was to play an important part both technically and politically at Itek and Perkin-Elmer.[19]

An interim report on the Perkin-Elmer Ad Hoc study was submitted to the CIA on 28 September 1964. Delivery of the final report which was originally scheduled for that date was delivered on 16 November 1964. The study contract was closed-out.[20]

On 28 September 1964, the CIA awarded Perkin-Elmer a four-month study to undertake the work proposed by Perkin-Elmer in their statement of work (MW-C-3).[21] Essentially, Perkin-Elmer was to select the single most desirable system for detailed Phase I study and design by 19 October 1964 and proceed with detail drawings suitable to produce an engineering model and prepare a program plan to achieve a production run of 40 units at a rate of one per month, aimed at a first flight on 1 November 1966. All items specified in the work statement were to be completed by 1 February 1965.[22]

A message from the CIA to Perkin-Elmer also mentioned that the camera contractor was to be selected by 1 March 1965 and revealed that a spacecraft contractor would be selected in early November 1965.[23]

The Perkin-Elmer program manager selected for this new four-month effort was Milton Rosenau. However, throughout the entire life of this program until his retirement in 1976, Earle Brown, a brilliant optical instrument engineer, was a major contributor of ideas. He produced a substantial amount of engineering work not only on the pioneer effort of the Hexagon program, but also participated in the final proposal and the design of the Hexagon camera. After his retirement, Earle worked for Perkin-Elmer as a consultant until his death in 1981.

Earle Brown, a Perkin-Elmer Staff Engineer, led the first AD HOC studies that were conducted at Perkin-Elmer on the Hexagon Program.

Work on the new study moved at a fast pace. In 12 weeks, over 22,500 manhours of engineering and experimental work were expended. Almost 1400 hours was unpaid overtime — effort of over 40 Perkin-Elmer people dedicated to this important program.

In support of this study, 30 supplementary engineering reports were prepared, and 16 technical memoranda and 10 breadboard test reports were written.

Preliminary parameters for the photographic system proposed in this study were established during Phase O. It was shown in the course of that effort that a system which would reliably resolve ground targets in the 2 to 5-feet range should have a focal length of approximately 80 inches and a relative aperture of f/4. These values were adopted from the system selections study which initiated the Phase I effort and which is reported in Supplementary Engineering Report No. 101 prepared by Earle Brown.[24] After consideration of a number of possible approaches, a Maksutov-type catadioptric system was selected as best meeting these requirements.

During the course of the Phase I Ad Hoc study, John Crowley and other program office personnel visited Perkin-Elmer. A formal presentation was given by Perkin-Elmer management and engineering personnel on 16 December 1964. Both technical progress and a discussion on program schedule and costs were presented.[25]

The importance of Perkin-Elmer's work was emphasized in mid-January 1965. In a message from Milton Rosenau to John Crowley, the Agency was informed that all contractual required documentation would be completed on the Phase I effort on 1 February 1965. Rosenau also requested permission to disassemble the breadboards, reschedule program personnel to other activities, and terminate all leases for equipment and contracts for security guard service.[26] John Crowley's reply was that the disassembly of the breadboards and the reassignment of personnel and termination of existing leases and contracts should be delayed.[27]

The CIA was in the process of preparing a briefing for the Land Panel on 23 February 1965. The objectives of the meeting, as stated by John McCone (DCI), was to examine the feasibility of the Fulcrum system to meet the requirements of the next generation search/surveillance system.

The Fulcrum program was presented by the Special Projects Office and its proposed contractor team: Sensor (Itek), Spacecraft (GE), Reentry Vehicle (AVCO), and System Engineering and Assembly (TRW).

Subsequent to the presentation and while the Land Panel was in caucus, Itek management, Frank Lindsay and John Wolfe, requested a private audience. They then stated that Itek was withdrawing from the Fulcrum Program and would not entertain a contract with the Agency to produce the Sensor.

Perkin-Elmer would shortly become aware of Itek's withdrawal from the CIA program. During a luncheon conference at a local seafood restaurant, Bob Sorensen, Ken Macleish, and Dick Werner (a program manager who had just joined Perkin-Elmer) were in the process of ordering lunch when Macleish received a phone call from Jack Maxey.[28] Maxey was calling from Bradley Field near Hartford, Ct. and asked how soon Sorensen and Macleish could get to the airport. Macleish said they would leave in two minutes and then informed Sorensen. After apologizing to Werner for stranding him, Sorensen and Macleish headed for the airport.

Arriving at the airport in record time, Macleish and Sorensen met with Jack Maxey and Les Dirks in a small conference room in the airport terminal.

Les Dirks asked if Perkin-Elmer could step up their effort on the Fulcrum program and assign Milt Rosenau as the program manager. Sorensen replied, "Yes and yes - unequivocally." The CIA representatives did not explain why there was a change of direction, only that it was a matter of great urgency. Les Dirks then asked Sorensen and Macleish to meet them in Washington in a day or two to discuss the details of the accelerated effort.

Shortly after the Bradley Field meeting, Dr. Wheelon appeared in Chester

Nimitz's office and asked if Perkin-Elmer could take over the CIA space reconnaissance program started by Itek. Nimitz accepted the challenge.

During discussions in Washington among Milt Rosenau, Charlie Hall, ▌▌▌▌▌▌▌▌▌▌▌▌▌▌▌▌ plans were made to extend Perkin-Elmer's involvement in the program through the month of February. Additional tasks were outlined in a message from Perkin-Elmer to the Agency. In addition, Milt reiterated that the data from the two contracts awarded during the Phase I study would be delivered to the Agency on 1 February 1965 as scheduled.[29]

Shortly after Wheelon's visit to Chester Nimitz, Jr., a message was sent by the Agency (code name ▌▌▌ to the other associate contractors on the program including the General Electric Company ▌▌▌ and SEAC ▌▌▌ The message read, "▌▌▌ (Perkin-Elmer's code name) is currently reviewing ▌▌▌ (Itek Company) camera and it is therefore incumbent upon the project office and the other associate contractors to convey to them complete details of all aspects of the Fulcrum Program. In line with this concept, AVCO (Reentry vehicle contractor) and ▌▌▌ project managers and engineering managers will arrive at ▌▌▌ on the morning of 16 March to present respective systems briefings. Of particular interest will be status of interface of S/C (spacecraft) and R/V (reentry vehicle) with payload (camera), but any area of mutual technical interest may be discussed. AVCO representives will be present at the detailed briefing by RCA representatives on their sheet-fed handling system to be presented at ▌▌▌ on 17 March."[30]

The GE representative who briefed the Perkin-Elmer team at the meeting was Paul Petty, then Director of Engineering and System Engineer for the GE Fulcrum Program. Petty was later to play a key role in the development of the Hexagon sensor at Perkin-Elmer as P-1 Model Director, Hexagon Program Manager, Deputy General Manager Operations, and then Vice President, General Manager of OTD.

The CIA was now fully committed to

Paul E. Petty was Director of the first flight (SV-1) unit and later became Vice President, General Manager of the Optical Technology Division.

work with Perkin-Elmer. During a visit to CIA project headquarters, Perkin-Elmer management was furnished with Itek reports.[31] Soon after, a group of Perkin-Elmer technicians and administrative personnel made several secret trips in trucks to the Itek Company. They returned loaded with the large optical bar brassboard, test and handling equipment, optical glass, optical fixtures, and a large quantity of reports and records. This material was transferred to a special wooden platform which had been constructed in the parking lot of the 50 Danbury Road Facility in Wilton, Conn. and then moved to the project area on the second floor.[32]

Perkin-Elmer was now embarked on a new phase of the Fulcrum Program (code name Protem at Perkin-Elmer) which would eventually culminate in a presentation in Washington, D.C. to the Land Panel (subcommittee on reconnaissance programs). This committee would review the Perkin-Elmer configurations and other competing systems and make a recommendation to the ExComm on continuing activity in this area.[33]

In the following weeks, Perkin-Elmer became more deeply involved with the other associate contractors. On 1 April 1965, a briefing was held to permit the various contractors to get acquainted, to discuss roles and responsibilities, start initial team tasks, and to discuss methods of program control. One of the revelations of this meeting was that the SEAC contractor (TRW-STL) would serve a dual function. First as a CIA Systems Engineering consultant not only participating technically but also reporting the progress of the other associate contractors; and second, as the Assembly Checkout contractor. (SEAC)[34]

This was a dual function which SEAC served during Itek's tenure as the camera contractor for the agency, and was an arrangement to which Itek objected.[35] Some feel it was a factor in Itek's decision to withdraw from competition since Itek was reluctant to allow another company to gain access to its operations and techniques.[36]

Perkin-Elmer was now fully immersed in the program. However, a message which Perkin-Elmer received from the agency, 13 April 1965, indicated that Perkin-Elmer's proposed designs were to be compared with proposals from other companies.[37] In fact, in that same message, the CIA project office informed Perkin-Elmer that, "This office has come to the considered judgement that the transfer on Fulcrum project information which has taken place is the maximum which the government can furnish consistent with the preservation of a competitive environment. Effective immediately, detailed interactions with the government on Fulcrum will be enjoined and limited to the issuance of a written statement of system and procurement requirements which will be issued promptly. We will of course entertain written requests for clarification of that document."

In that same message, Perkin-Elmer was instructed to submit a formal proposal of their selected designs by 7 May 1965. The Agency was apparently preparing for the upcoming "Land Panel" meeting which was initially planned for 25 June 1965. The committee was to review the Fulcrum, Matchbox, and other competing configurations. The message also noted that Perkin-Elmer should position itself for possible program termination on 30 June 1965.

On 4 May 1965, a preliminary presentation was given by Perkin-Elmer management to Dr. Wheelon at CIA headquarters. He was briefed on the two quite different systems which were emerging from the Perkin-Elmer studies; a modified version (F') of the original Itek (F) design, and a modified version (M') of Perkin-Elmer's Matchbox (M) configuration.[38] The formal technical presentation at the agency was given on 19 May 1965.[39]

It was during the course of this activity that a change in the managerial direction of the Fulcrum program at Perkin-Elmer began to occur. After completing the Ad Hoc Phase I study program and preparing the proposal for the 7 May 1965 presentation, Milton Rosenau continued to lead the engineering effort on the program reporting to Richard Babish, and through him to Dr. Macleish. However, in order to place more emphasis on this activity, the management of the program was changed from Rosenau to Macleish.

The period from the beginning of March 1965 to the end of July 1965 was very active. In addition to studying the Itek (F) system, Perkin-Elmer examined other systems. The customer favored the "F" system concept. However, Perkin-Elmer was reluctant to follow through on that design since it had many shortcomings. It did, however, have one significant advantage. As mentioned previously, one of the critical problems of space camera systems is thermal control. The "optical bar" design of the "F" system was an answer to that problem. The "optical bar" was first noted in an engineering notebook of an Itek engineer in preparation for Itek Proposal No. 3233.[40] The optical bar not only helped to maintain thermal control of the optical system, but also provided the scanning motion. Unfortunately, the Itek film transport design proved to be unwieldy and cumbersome. The optical bar also required a large perforated optical flat. The manufacturability of the flat had not as yet been demonstrated by Itek.

Although the customer realized the shortcomings of the "F" system, it still favored the "optical bar" concept. An early advocate of the "optical bar" at Perkin-Elmer was Dr. Rod Scott. In addition, Robert M. Landsman, a young and competent electrical engineer recommended a "one-pass" modified F' design[41] and Milton Rosenau wrote a two-page report proposing that Perkin-Elmer recommend the "optical bar" design with a "one-pass" film transport system.[42] Perkin-Elmer management, however, felt that a more thorough study was required before a final decision could be made.

The Land Panel meeting was moved up a month to 21 July. In preparation for this meeting, a dry run attended by the customer, Perkin-Elmer, General Electric, and SEAC representatives, was held on 20 July.[43]

After the Land Panel meeting, the CIA sent a message instructing Perkin-Elmer to embark on a program definition and final configuration study phase which would last three months.[44] The Project Office requested Perkin-Elmer to select a final configuration by 18 August 1965.[45] A modified three-month work statement later changed this requirement to 14 September 1965.

In a summary report, Perkin-Elmer documented the principal considerations entering into their recommendation at the F' vs M' briefing at CIA headquarters on 14 September 1965. The recommendation was to pursue the F' system and discontinue further work on the M' system.[46] A message from the Agency dated 17 September 1965 instructed Perkin-Elmer to terminate all work on the M' system.[47]

This conclusion was the result of a long and careful examination of Itek and Perkin-Elmer systems which Perkin-Elmer began on 1 March 1965. It was a difficult decision since some of the participants at Perkin-Elmer were opposed to the Itek system under any circumstances, while advocates of the modified Itek (F') system argued that Perkin-Elmer improvements made it a feasible design. The major improvements to the F' system were: (1) a practical method of transporting the film on and off the rotating "Bar", and (2) a method of manufacturing the flat.

Perkin-Elmer, the Project Office, and the associate contractors could now concentrate all their efforts on one configuration, the F' system. Dr. Macleish, who directed the program personally since May 1965, decided that this was an opportune time to reorganize the Engineering Department so that more attention could be concentrated on the Fulcrum Program. He placed W. Richard Werner in charge of the Special Projects Branch. The Ad Hoc Group was part of that branch and reported to W. Richard Werner.[48]

The new study program, which was started on 1 August 1965, required a reporting system. On 27 August 1965, the first message (Biweekly TWX Report), prepared in accordance with Task 14 of the three-month study work statement, was sent to the customer.[49] Shortly after the 12 September meeting in which the F' system was selected, Richard Babish, who prepared the first three biweekly TWX reports, started a new series of progress reports in accordance with Task 9 of a revised work statement dated 21 September 1965.[50] From that point to the present, there has been an unbroken chain of reports of technical progress, program costs, and project schedules. As a matter of interest, the first TWX report indicated that 58 Perkin-Elmer people were assigned to the program on the week ending 13 August 1965.

The new study program was scheduled for completion on 15 November 1965. It ultimately lead to a summary report and a presentation to an NRO task group by Perkin-Elmer on 9 December 1965.[51] The presentation defined the study objectives and the results of the three-month effort which included (1) the preparation of a detailed proposal describing the design of the proposed system (2) the accomplishment of a series of experiments selected to explore and solve key problems and (3) the interchange of technical findings among the associate contractors in the interests of devising an integral search/surveillance system.

A final report which summarized the

work accomplished during the three-month period was submitted to the customer on 28 February 1966.[52] It included a plan for a six-week period for the proposal presentation which would commence with the receipt of the formal request for a proposal (RFP) from the Government. The plan included the formation of a division within the newly organized Optical Group formed in June 1965 and headed by Robert H. Sorensen. The new division responsible for the program would be called the Optical Technology Division (OTD) and would be managed by W. Richard Werner. The Ad Hoc Project, formed in June 1964 on the inital study contract, was to be elevated to department level within the division, headed by Michael Maguire.

Perkin-Elmer was now gearing for the final phase of the Fulcrum program, preparation for the final request for proposal (RFP). As time went on, program personnel started getting anxious. An internal progress report dated 16 February 1966 stated, "The program office (customer) does not have any additional information to support the date that an RFP will be forwarded or contract awarded."[53] But the day finally arrived. On 19 May 1966 Perkin-Elmer received an RFP from the Government requesting that a proposal be submitted by 21 July 1966. The 156 Perkin-Elmer people assigned to the program at that time began a grueling effort that would entail working overtime and spending weekends finalizing reports and completing experiments.

From that summer day in June 1964, when Les Dirks and Jack Maxey first approached Perkin-Elmer, to the day the Government issued the RFP, Perkin-Elmer's modest involvement in the program changed to a company positioned to make a significant contribution to our national defense.

Michael Maguire managed the AD HOC Department in 1965 and eventually became Vice President, General Manager, Optical Group East and Optical Technology Division. His engineering leadership was the driving force on the Hexagon Program at Perkin-Elmer.

ORGANIZATIONAL PERIOD

At the time that Perkin-Elmer first became involved in the CIA reconnaissance study (Fulcrum), later to become the Hexagon Program, the company was under the leadership of Richard S. Perkin, Chairman of the Board and co-founder of the company. Robert E. Lewis was President of the company and Chester W. Nimitz, Jr. had just been promoted to Executive Vice President of Operations (elected President and chief executive officer on 1 January 1965). Robert H. Sorensen was Vice President and General Manager of the Electro-Optical Division, Dr. Roderic M. Scott was Vice President and Chief Scientist, and Dr. Kenneth G. Macleish was Vice President and Director of Engineering of the Electro-Optical Division.[1] These were the major participants in a series of events that would have an impact on a critical national defense program and the future of Perkin-Elmer.

Just eight years prior to the start of the Ad Hoc project, the company separated its functions into a commercial division and a government division. The Electro-Optical Division, under the direction of Rod Scott, developed a strong base of experience on government programs and nurtured a creative engineering force. In June 1965, this area of company business was elevated to Group status, when the Electro-Optical Division and the newly established Astro-Optical Division became part of the Electro-Optical Group.[2] The new Group was headed by Robert H. Sorensen, Group Vice President, who also continued as General Manager of the Electro-Optical Division.

Soon after Dr. Macleish joined the company in January, 1962, he reorganized the Engineering Department in the Electro-Optical Division into several branches. Branch D. was managed by Richard C. Babish and included a Special Projects Section responsible for classified reconnaissance programs. Milton D. Rosenau was in charge of the Engineering Department reporting to Richard C. Babish.[3] The project team for the camera study, which started on June 23, 1964, was formed from personnel in this section.[4]

Earle Brown, a Staff Engineer reporting to Dr. Macleish, was appointed project manager of the initial three-month study. He was guided by a committee consisting of Dick Babish, Rod Scott, Milt Rosenau, Bob Hufnagel, and Kenneth Macleish.

Perkin-Elmer received additional contracts on September 29, 1964, for follow-on work to the original three-month study. The new four-month effort was directed by Milton D. Rosenau. The engi-

Perkin-Elmer Corporate Headquarters, Norwalk, CT (June 1964)

neering force was expanded to 40 people, also from the Special Projects Section. A final report on the four-month study includes an organization chart listing these people.[5]

At the end of this activity, in February, 1965, the CIA contracted with Perkin-Elmer for an additional month to keep the project team together and to work on particular problems. This effort was soon expanded by additional contracts which remained in force until the award of the Hexagon Program to Perkin-Elmer on October 10, 1966.[6]

In the period from June to December, 1965, Dr. Macleish directed the program with the assistance of Richard C. Babish and W.R. Werner, who was being prepared to take over program direction in December.

On December 8, 1965, a day before an important high-level presentation to the customer, Kenneth Macleish established the Ad Hoc Program Department, originally called the Ad Hoc Project.[7] By this time, the name of the Engineering Department, headed by Dr. Macleish, had been changed to Program Operations, and the letter designations of the branches were changed to more descriptive titles. The Ad Hoc Group, which was elevated to department status, was to be managed by Michael Maguire who previously managed the Systems Department in Program Operations.

W.R. Werner was made responsible for the line management of both the Special Instruments Department and the new Ad Hoc Department, reporting to Kenneth Macleish. Milton Rosenau was made Manager of the Advanced Development Section in the Special Instruments Department.

The December 9, 1965 presentation included a discussion on a proposed project organization establishing the Optical Technology Division. Although there were changes in the lower management levels, the basic project structure remained unchanged through the proposal phase.[8] A new organization chart appeared in the proposal which Perkin-Elmer submitted to the government on July 21, 1966.[9] Since then the OTD organization has been fine-tuned to respond to the needs and priorities of the Hexagon program. As in any activity, the needs of the project and the talents and capabilities of the people associated with the project usually dictate the manner in which an activity is organized. (See Appendix H.)

Dr. Roderic Scott (left), Chief Scientist in 1964, managed the early aircraft reconnaissance systems at Perkin-Elmer and made major design contributions to the Hexagon Sensor Subsystem. Dr. Kenneth G. Macleish, Vice President and Director of Engineering, Electro-Optical Division in 1964, managed the early engineering studies on the Hexagon Program.

Richard S. Perkin (left), Chairman of the Board and Co-founder of the company, and Chester W. Nimitz, Jr., Executive Vice President of Operations, headed the company during Perkin-Elmer's entry into space reconnaissance activities in 1964.

EARLY TECHNICAL DEVELOPMENT

The development of the U.S. fourth generation photographic space reconnaissance camera started in the spring of 1964. Preliminary work began at the Itek Corporation, Burlington, Massachusetts on an information gathering system which could achieve the ground resolution necessary for spotting reconnaissance, while at the same time, provide the wide coverage needed to carry out a search mission from an orbiting satellite.

The parameters affecting microwave, infrared, and visual spectrum band sensors and the resolution capabilities of these devices were examined. The conclusion was reached that an optical system working in the visible region of the electromagnetic spectrum was the best choice.

After making the basic selection, Itek submitted a proposal in May 1964 recommending a panoramic type camera with a capability of providing high ground resolution and wide area coverage. The configuration selected was termed the "optical bar."[1] The arrangement is basically a reflective system with refractive elements used only to correct aberrations and flatten the focal plane. A scanning or panoramic type of camera was selected since it allows use of a high quality optical system with its inherently narrow field in a broad coverage system.

This type of system has the advantage of a large scanning angle limited only by vehicle window size or vignetting by an adjacent camera used in stereo coverage. Tradeoff studies involving depth of focus, exposure time for Eastman Kodak Type 4404 film, available vehicle space and weight limitations resulted in the selection of a 60-inch focal length, f/3.0 optical system. The newly emerging science of using Modulation Transfer Functions for the selection of system parameters and film characteristics guided the selection of focal length and aperture size.

It was at this point that the CIA decided to fund a parallel study (Fulcrum) at Perkin-Elmer as a backup to the work being performed at Itek. This was not unusual since the government frequently has several companies working on the same problem to insure the development of the most effective system.

On 23 June 1964, Perkin-Elmer started work on a three-month study of a satellite-borne photographic reconnaissance camera with the following operational requirements.

Ground resolution:	2-5 feet
Operating Altitude:	100 nautical miles
Coverage with stereo:	Continuous swath, approximately 200 nm wide

The system was to use Eastman Kodak Type 4404 film with suitable allowance for potential improvement in exposure index. Target contrasts, as observed from the aperture of the camera, were to be 2:1. Additional requirements mentioned in the final report was a space restriction imposed by a 10-foot diameter vehicle size, a 30° stereo convergence angle and a total scan angle of 90°.

The attention of the Perkin-Elmer study was directed toward definition of the optimum form or forms of systems, power and environmental requirements, compatibility with film recovery systems, control methods, operational reliability, and the identification of special problem areas.

A parametric study resulted in the selection of a photographic system with an aperture of 20 inches, a focal length of 120 inches, covering a 9-inch film format and scanning a 200-nautical mile swath, capable of a ground resolution of 2 feet at nadir. This set of parameters represented an upper limit of system size, weight, and performance level and was used as the basis for the investigation of alternate configuration concepts.

Of the possible camera types, the frame camera and the strip camera were eliminated as candiates in the early stages of the study and the effort was concentrated on developing configurations of various panoramic cameras.

At the outset, it was apparent that an optical design of high performance would be required to meet the specifications. Study of previous work by

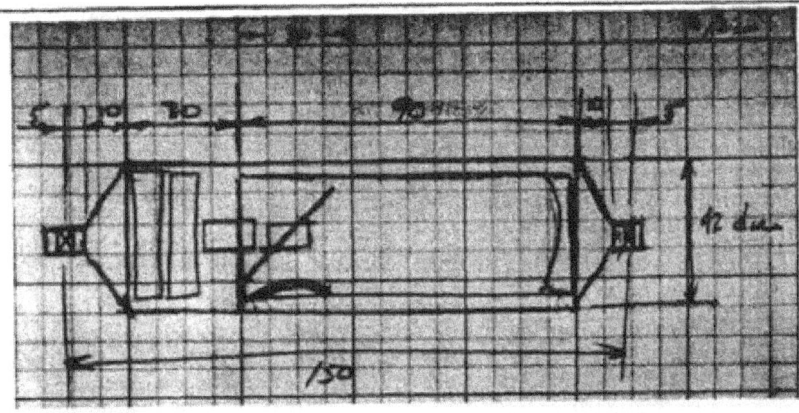

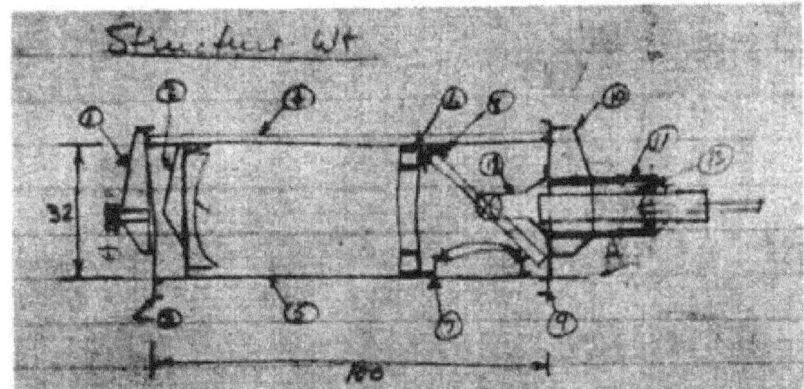

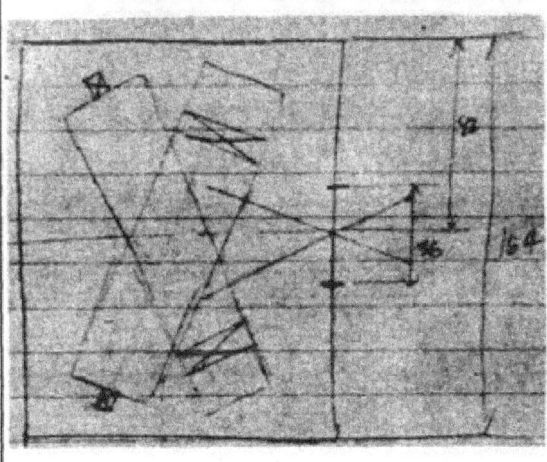

Early sketches of the "Optical Bar" taken from an Itek engineering notebook dated 22 May 1964.

The top sketch shows the primary mirror (right side), the diagonal mirror and the corrector plate (center), and an outline of the supply and take-up film spool (left).

The center sketch identifies the various parts of the Optical Bar. Major parts include the primary mirror assembly (2), the diagonal mirror (8), and the field lens assembly (12).

The lower sketch shows the spacecraft envelope containing two optical bars.

Perkin-Elmer of large near-diffraction-limited optical designs suggested the further study of three types for detailed comparison: the Petzval, a folded Maksutov, and the Flat Field Schmidt. Of the three designs considered, the folded Maksutov and an unfolded variation were the most suitable for use in the system.

It also became apparent in the course of the study that the incorporation of a transfer lens in the system would provide substantial advantages in nearly every possible configuration. The basic choice of an optical arrangement was a catadioptric system consisting of an objective mirror and a correction lens at the entrance pupil. A transfer lens system has the function of transferring the image from the focal plane of the primary optical system to a second focal plane in which the film is positioned.

A large number of possible configurations were studied and included factors such as method of scanning and film transport techniques. One system characteristic given considerable weight in this investigation was the capability of extracting film from the supply at a constant rate during the photographic period. The Perkin-Elmer study resulted in three alternate configurations: Matchbox, Scarecrow, and Ferris Wheel.

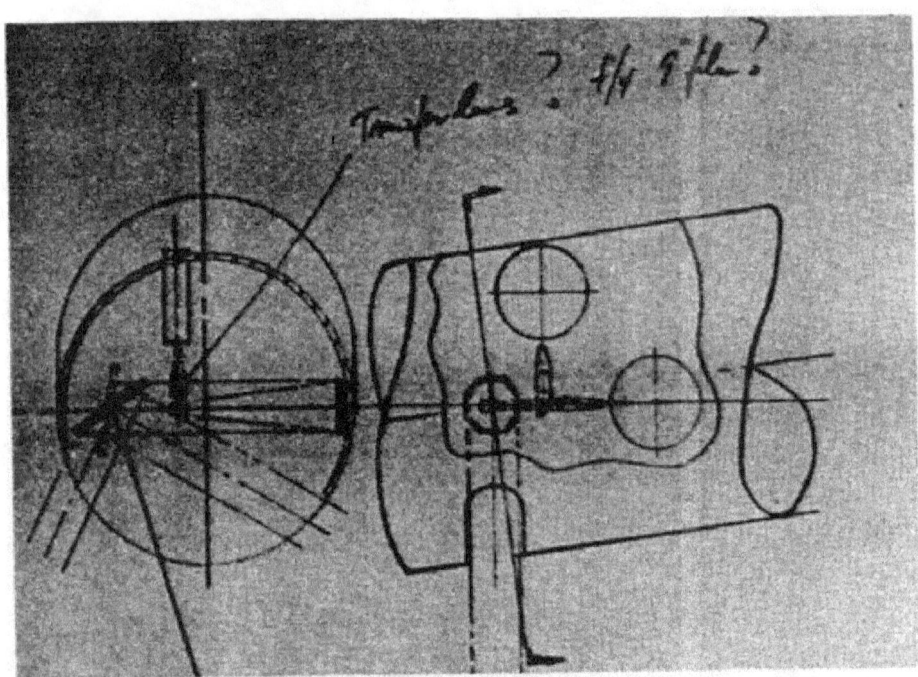

Matchbox: This system scans with an oscillating mirror. Its optical system, film supply, and take-up are fixed in a stabilized vehicle. It provides two independent systems which can be operated separately in the event of failure of one. Estimated weight of this system was 2800 pounds.

Focal length	120 in.
Aperture diameter	20 in.
Duty cycle	1.8
Film velocity	112 ips

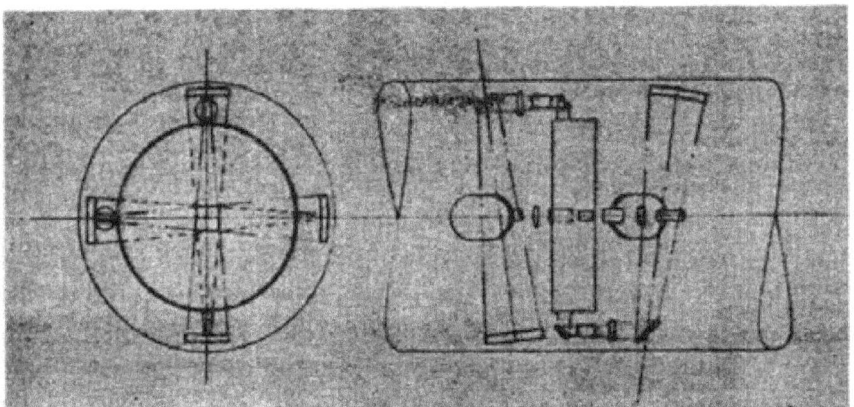

Scarecrow: This system scans by rotating a section of the vehicle, which includes the film take-up and the reentry vehicle. Film transport is over a counter-rotating drum. The estimated weight of this single system configuration was 3150 pounds.	Focal length Aperture diameter Duty cycle Film velocity	120 in. 20 in. 1 201 ips

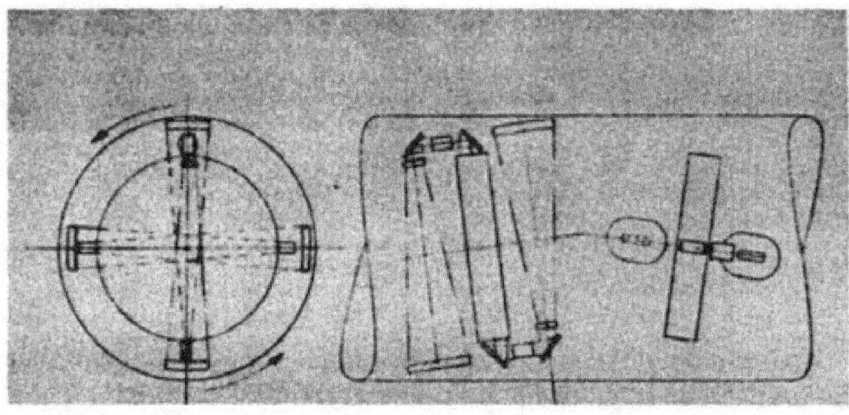

Ferris Wheel: This system scans by rotating lenses. The film supply and the take-up are fixed in a stabilized vehicle. Film transport is over a counter-rotating drum. It provides two independent systems. Total estimated weight of this system was 3060 pounds.	Focal length Aperture diameter Duty cycle Film velocity	90 in. 20 in. 1 112 ips

The study report was to be completed by 28 September 1964. However, work did not proceed at the anticipated pace and as a result, no sound technical basis had been established for the choice of one of the alternate systems.

At a meeting on 14 September 1964, the CIA asked Perkin-Elmer to submit a proposal for three-months additional work on the initial study, including a preliminary design of all three configurations selected, plus some component development and experimental work. However between that meeting and 25 September, Perkin-Elmer received new instructions.

Perkin-Elmer was to select the single most desirable system for detailed Phase I study and design by 2 October 1964. In addition, the work statement included a requirement for a program plan that would result in a production run of 40 units at a rate of one per month, aimed at a first flight on 1 November 1966. This was exactly the same activity that Itek began on 1 September 1964. Two other companies were involved on this effort. The General Electric Company was developing a spacecraft and the Avco Company was working on a reentry vehicle for the proposed camera system. Perkin-Elmer was now in direct competition with the Itek Corporation for a reconnaissance system.

Milton Rosenau was selected as Program Manager on the new study program. The project team was increased from the initial group of approximately 12 people to over 40 engineers and technicians.

The CIA established preliminary Fulcrum camera performance specifications which included the following:

> The camera payload was to be launched in a Titan II vehicle. It was to be an area coverage (search) system with the best possible resolution and include continuous stereo coverage with equal quality pairs at 30° convergence angle. The system was to be capable of monoscopic operation (single camera) using Eastman Kodak Type 4404 film. The maximum system weight was 2200 pounds with a film supply not exceeding 900 pounds (a requirement arrived at on the basis of a 10-day mission lifetime). The ground resolution at nadir was to be better than 5 feet, and the scan angle specified for the system was 120°. It was also desirable to obtain the maximum possible area coverage per mission, although no firm minimum coverage requirement was specified. The specification also noted that the scan angle (120°) and the stereo angle (30°) were under study and might be revised downward. The specification also revealed that the proposed design had to be compatible with a late 1966 first launch, given a March 1965 program approval. The spacecraft contractor was to be selected by early November 1964.

The parameters for the optical system established in the initial three-month study (Phase 0) were the basis for the Phase I Study. It was shown in the course of the Phase 0 effort that a system which would reliably resolve ground targets in the two to five foot range should have a focal length of approximately 80 inches and a relative aperture of the order of f/4. These values were adopted in the system selection study which initiated the Phase I effort and were reported in the Ad Hoc Supplementary Report 101.[2] This was a continuation of Earle Brown's study in the Phase 0 effort.

The Earle Brown report provided a classification of panoramic camera systems with the characteristics required by the Fulcrum specifications. The classification identified 576 possible system arrangements, some of which were illogical, but some of which could be realized in several variations of form.

Sixty systems were selected for consideration in choosing the optimum configuration. A Maksutov optical system with a relative aperture of f/4 was selected. Refractive systems were not considered feasible for the Fulcrum application and the Maksutov system was sufficiently representative of catadioptric systems for the purpose of the study.

First stage elimination reduced

considerations to five categories: Oscillating Turnstile, Rotating Turnstile, Oscillating Broomstick, Rotating Broomstick, and the Rotating Ferris Wheel. After consideration of components needed in the various forms (a process in which the need for a large flat mirror was a significant factor), the formal choice eventually fell to an Oscillating Turnstile system as possessing the most advantages and the least disadvantages.

The following are the system specifications of the Fulcrum system:[3]

Focal Length	72 inches
Aperture	f/4
Speed	T/7.9
Camera Weight	1700 pounds
Film Weight	1100 pounds (34,000 feet x 8.0 inch)
Coverage	
-Stereo	5.8×10^6 nautical square miles
-Mono	11.6×10^6 nautical square miles
Payload Size	114 inch diameter x 132 inches long
Exposure Time	1/200 second on 4404 film (30° sun angle)
Slit Width	0.315 inch
Film Velocity	63 inches/second at V/h = 0.042 rad/sec
Resolution at nadir	2 to 3 feet
Stereo Angle	30°
Scan Angle	+45 (90° total)
Orbit Altitude	100 nautical miles

While the fundamental design approaches adopted on the Phase I program represented techniques which were proven effective in prior hardware development efforts, there were a number of areas in which it was deemed advisable to design and construct breadboards to prove feasibility of the design approaches; to make quantitative measurements to assure meeting the camera specifications; and to more effectively visualize the system in a three-dimensional form so that structural design and space utilization could be optimized. Seven design areas were breadboarded and tested:

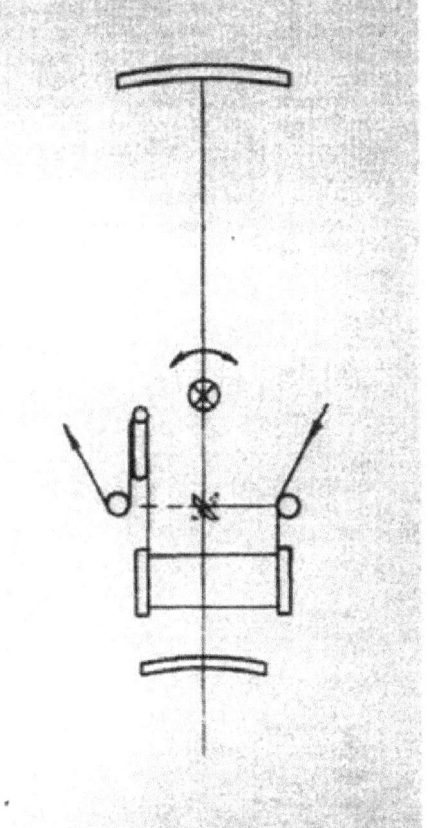

The Turnstile configuration has a roller film transport at the focal plane. The fold is taken in a plane normal to the slit, and the oscillation of the film transport is around an axis parallel to the film plane, which avoids the necessity for twisters.

(1) The film transport system, from supply through the camera to takeup, was breadboarded in full scale (this was called the "cocktail shaker").

(2) A small optical mockup was constructed to demonstrate the correctness of the image orientation and motion analysis.

(3) Two structural mockups of the complete system were made; a

Hx TOP SECRET

small-scale "Meccano" mockup to aid in space visualization, scan cycling techniques, and structural concepts, and a full-scale lens cone mockup (cocktail shaker) to furnish the effective visualization of the overall system size and weight, to demonstrate the nature of the dynamic interactions involved, and to furnish a dynamic vehicle for the full-scale transport system mockup.

(4) An electromechanical functional breadboard of the essential elements of the film velocity control system, including the requirements for setting of the nominal velocity by a focal length adjustment "knob," and for modulation of the film velocity during scan, was constructed and tested.

(5) A dynamic functional model of the camera system was constructed to study the dynamics of the scan cycle, as an input to the effective design of a minimum energy camera drive control, was built.

(6) A small-scale mockup of the film transport drive mechanism and the exposure slit area was constructed and tested to aid in the development of the film transport control servo systems, and to demonstrate the effectiveness of the differential film tensioning device which maintains adequate tension in the platen area without the expenditure of large quantities of energy at the supply and takeup spools.

(7) A Ronchi Grating Interferometer was constructed to measure film flatness in the platen area to determine the tension requirement to maintain camera focus within the established tolerance.

The Perkin-Elmer four-month study report and proposal were submitted to the CIA on 1 February 1965 as scheduled.[4]

While Perkin-Elmer was working on the Fulcrum, Itek was involved in the feasibility study of a configuration ("optical bar" concept) they had proposed to the CIA in their May 1964 proposal. Itek started Phase I of the Fulcrum program on 1 September 1964 and constructed various breadboards and a full scale working model of the "optical bar."

Itek submitted a final report to the CIA on 26 February 1965, a month after Perkin-Elmer submitted their version of the Fulcrum system.[5] Unlike the Perkin-Elmer system which consisted of two lens cones using an oscillating scanning motion, Itek's "optical bar" scanned in a continuous motion. The basic idea of the "optical bar" was developed by Itek in a 1962 study. The configuration is known to most optical companies, however, it was Itek that first applied it to the Fulcrum system. The proposed design was a 60-inch focal length, f/3.0 optical system, with a transfer lens system. The following are the system specifications of the Itek Fulcrum design.

Focal Length	60 inches
Aperture	f/3.0
Speed	*
Camera Weight	2340 pounds
Film Weight	*
Coverage	
-Stereo	5.8×10^6 nautical miles^2
-Mono	11.6×10^6 nautical miles^2
Payload Size	*
Exposure Time	*
Slit Width	*
Film Velocity	*
Resolution at Nadir	*
Stereo Angle	*
Scan Angle	± 60 (120° total)
Orbit Altitude	100 nautical miles

*Not included in report.[5]

Itek's design was based on experience acquired during development and test of a full-scale film transport brassboard and critical optical elements.[6]

Initially, 1 March 1965 was the date

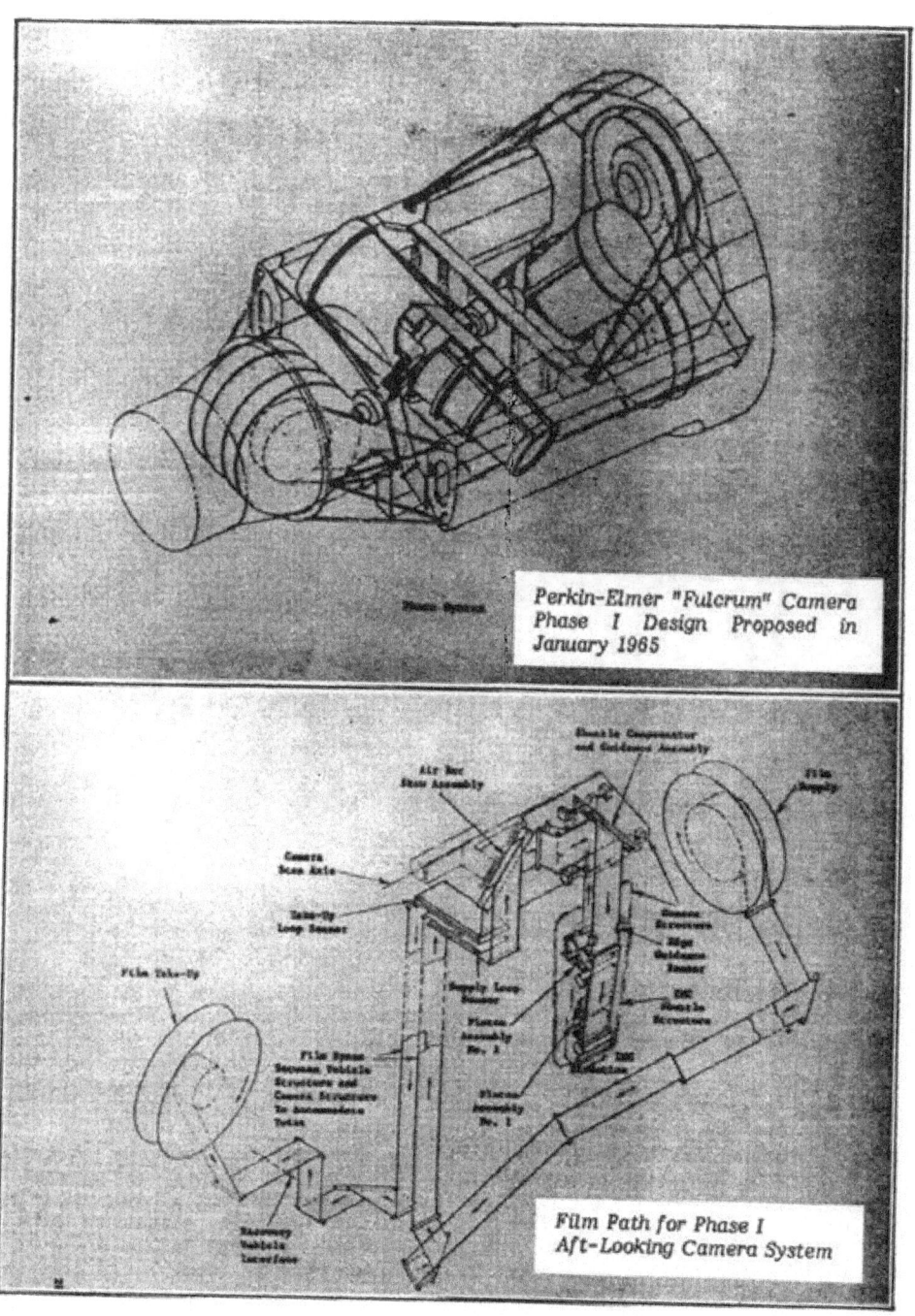

Perkin-Elmer "Fulcrum" Camera Phase I Design Proposed in January 1965

Film Path for Phase I Aft-Looking Camera System

Full scale mockup of Perkin-Elmer's Phase I design. Due to its configuration and operation, it was nicknamed the "cocktail shaker." It contained a film transport system consisting of rollers and air bars, and transferred film from a supply spool to a take-up spool.

on which the camera contractor was to be selected. However, just prior to the date, Itek decided to withdraw from the (CIA) competition.

The CIA was now placed in a dilemma. It favored Itek's concept of the "optical bar" over Perkin-Elmer's "cocktail shaker", but wanted the full 120° scan which Itek considered unnecessary since it would be used infrequently.

The CIA decided to ask Perkin-Elmer to continue the development of the "optical bar" configuration. On 1 March 1965, Bud Wheelon, head of the Science and Technology Group of the CIA, visited the Perkin-Elmer Corporate headquarters in Norwalk, Connecticut to discuss this possibility. Mr. Nimitz agreed to continue the program; however, he was reluctant to do so without a thorough analysis of the Itek system.

Perkin-Elmer engineers began their review of the Fulcrum or "F" system (also code named Protem at Perkin-Elmer) and after working on the analysis for almost 8 weeks, Perkin-Elmer developed a modified version of Itek's concept, the "F-Prime" system. In addition to the F' design,

Itek's Engineering Brassboard of the Optical Bar

Perkin-Elmer also developed an entirely different configuration called the M' system. A system called the Matchbox or "M" system had been studied during the Perkin-Elmer Phase "O" effort from June-September 1964. An "M" system design, funded by the USAF, was also developed by the Eastman Kodak Company. This study contract was transferred to Itek sometime after March 1965.

The concept was resurrected, and with modifications and improvements, was developed to fulfill the new requirements of the Fulcrum system. At the time that Perkin-Elmer developed the F' design, Macleish felt that it was an unnecessarily big, heavy, and expensive system and decided that it would be a good idea to look into a simpler system and compare the two systems (i.e. the F' system and the M' system (based on the Matchbox configuration). These were defined in a CIA message to Perkin-Elmer dated 14 April 1965.[7]

The new Fulcrum camera system requirements were to be the basis for evaluation of proposed designs which were scheduled to be presented to an NRO study panel. The specifications were expanded and refined but essentially contained the original Fulcrum requirements. The 14 April message specified that the camera scan angle had to be at least 90° with a scan angle capability up to 120° adjustable prior to launch. The camera system design weight ceiling was established at 3400 pounds, including film and camera mounting structure, but not ancillary camera hardware such as the Stellar/Index unit and the recovery takeup assembly.

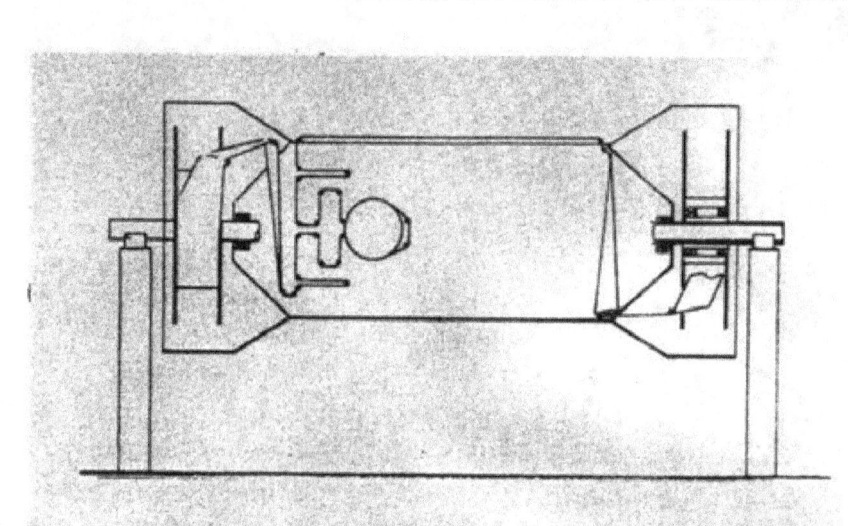

Profile of Itek's Brassboard Configuration

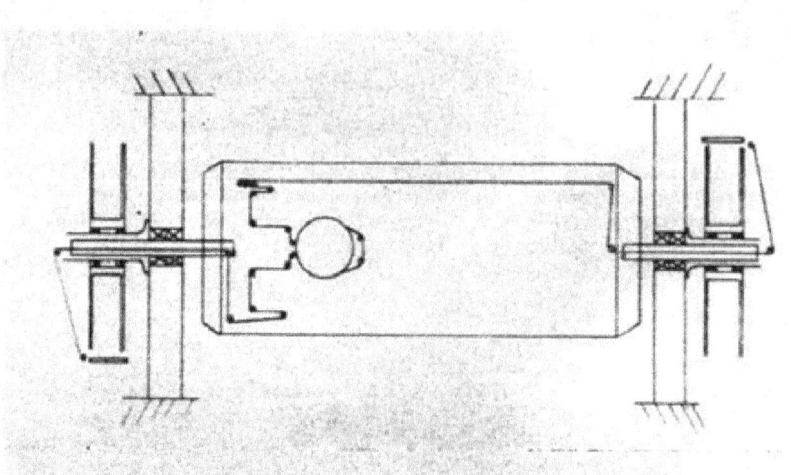

Profile of Itek's Camera Configuration

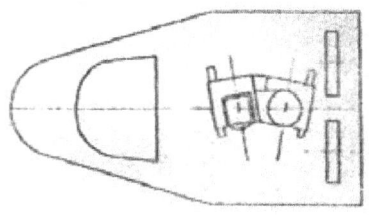
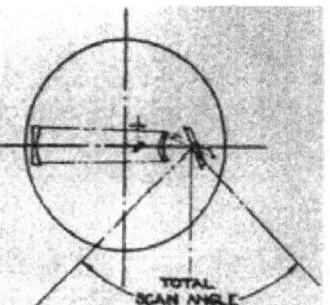

Outline of the Perkin-Elmer "M" Camera Configuration.

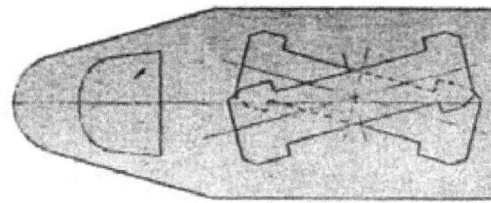

Outline of the Itek Fulcrum (F) Camera System.

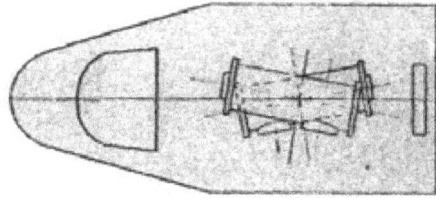

Outline of the Perkin-Elmer modified version (F') of Itek's (F) camera system.

Perkin-Elmer presented both the F' and the M' systems at the Land Panel briefing which was held on 21 July 1965. The purpose of the panel meeting was to review the CIA's proposed systems (F' and M') and other competing systems which included a new Itek design now sponsored by the USAF, and an Eastman Kodak concept (also funded by the USAF).

The Land Panel decided that a three-month system definition phase was necessary to firm up the designs of the competing systems. Perkin-Elmer was to make a final choice between the F' and the M' systems prior to 18 August 1965. On 14 September 1965, during a presentation to the CIA, Perkin-Elmer recommended dropping the M' system and continuing work on the F' system. This was approved by the CIA.[8]

A revised work statement was formulated by the CIA on 17 September 1965 to define the work that was to be accomplished on the F' system design. Various engineering and experimental tasks were outlined and completion dates were specified.

There was to be one more preliminary review of all competing systems by the National Reconnaissance Office prior to the release of the Request For Proposal (RFP) for the fourth generation reconnaissance camera. The Director of the NRO established a task group to provide him with information which would be used to assist him in

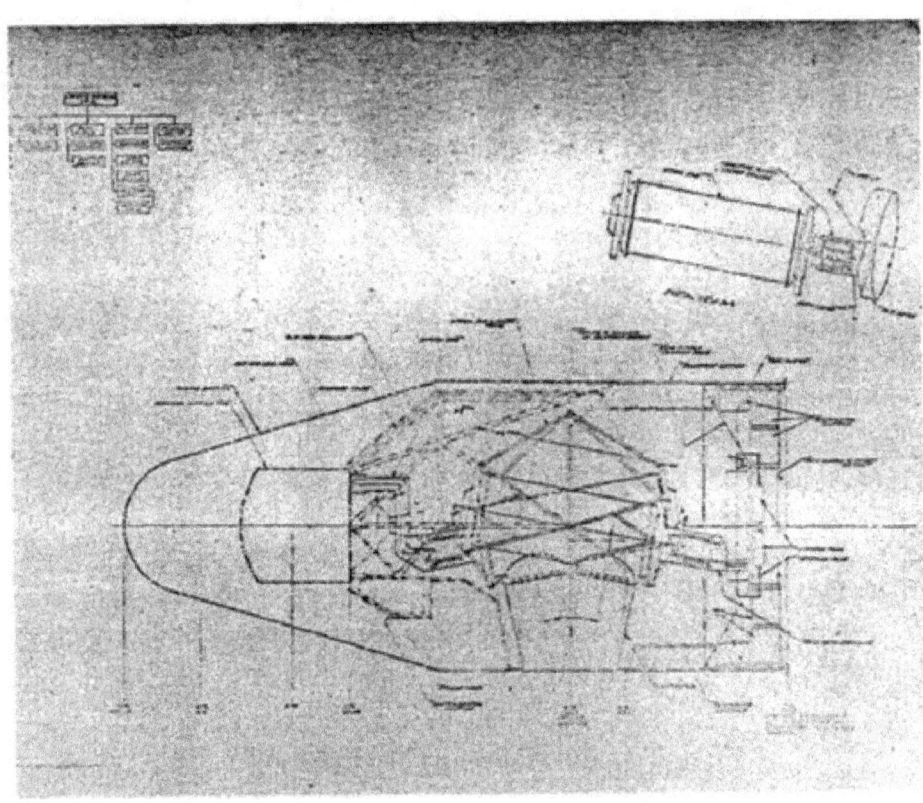

Initial Concept Drawing of Perkin-Elmer's Sensor Subsystem Design

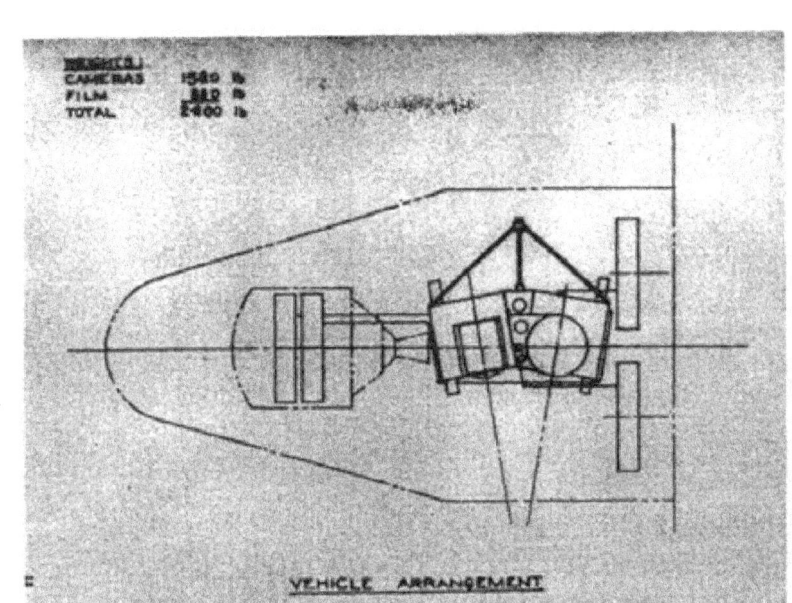

The Original Layout of Perkin-Elmer's M¹ Camera System

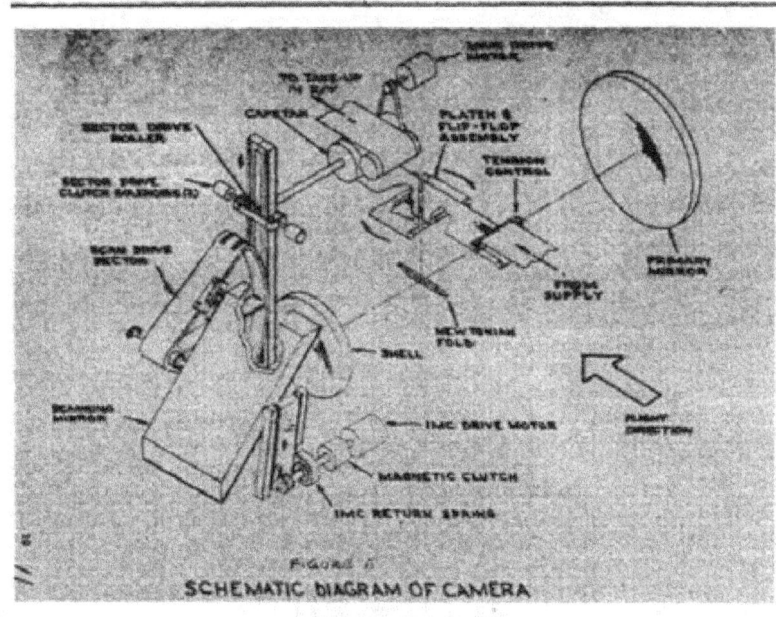

M¹ System Schematic

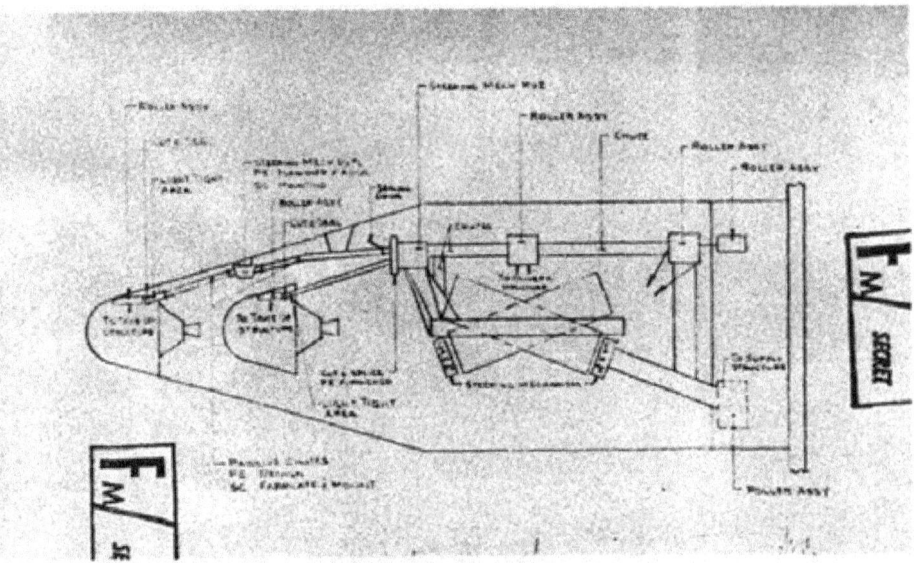

The Conceptual Drawing of Perkin-Elmer's Hexagon Camera System

determining the course of action at the camera contractor companies before and during source selection.[9] The task group was scheduled to visit Perkin-Elmer on 9 December 1965 and Itek on 10 December 1965. The Perkin-Elmer presentation included the status of the F' design activity, a brief historical account of the company's activities relevant to the new search/surveillance system, and the projected state of the project in the time span between 1 January 1966 and 1 April 1966.

After the presentation to the NRO representatives, Perkin-Elmer continued its work on the F' system and made preparations for the final proposal. The RFP for the fourth generation system was finally released to the camera contractors on 21 May 1965.[10]

For the next 8 weeks, the major effort at the newly formed Ad Hoc Department was proposal preparation. Engineering analyses and experimental tests were conducted to support the technical proposal.

The 127-page RFP contained not only the design criteria and performance requirements of the new reconnaissance camera, but also detailed requirements for program control and administrative functions. The proposal request included a further refinement of operational and performance requirements previously stated in funded studies and work statements.

The required ground resolution for the system from design perigee altitude was to be 2.7 feet or better at scan nadir. Stereo coverage of at least 20° but not greater than 30° was to be provided. The scan angle was to be at least ±45° but not to exceed ±60°. The required daily coverage was over 566,500 nautical square miles.

The specified ground resolution resulted from work performed by Frank Scott in Perkin-Elmer's Research Department. Working with Drs. Donald N. Buckner and Al Harabedian of Human Factors Research, through a subcontract, Scott conducted psychophysical assessments of photo-interpreter performance (not necessarily preference) as a function of ground resolution. A series of such studies were conducted for Les Dirks.

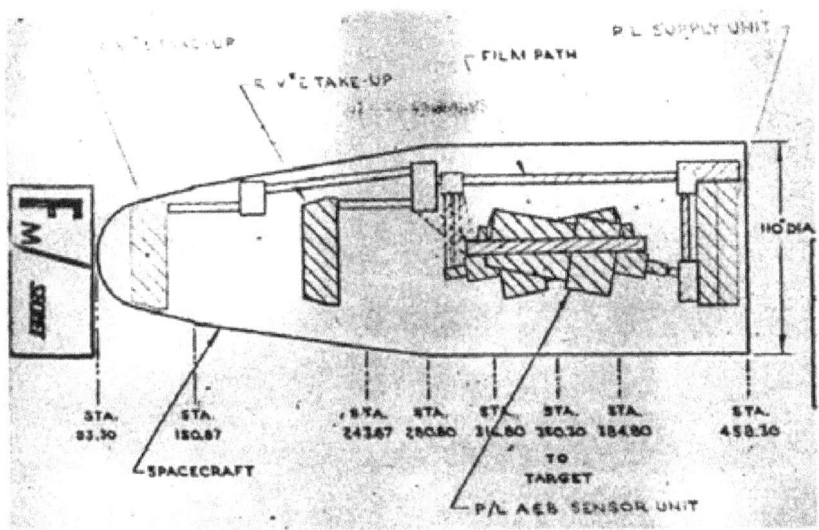

Early Perkin-Elmer Sketch of Hexagon Camera Showing Major Components

and ▓▓▓▓▓▓ of the National Photographic Interpretation Center assisted and made available practicing professional photo-interpreters to serve as subjects. Likewise, studies showed photo-interpreters performed best with stereo convergence angles of about 10° while photogrammetrists minimized mensuration errors with large angles, greater than 30°; to satisfy both kinds of users of the prospective Hexagon System images, Les Dirks decided on 20°.

The new camera was to be launched in a Titan IIID and capable of operating 40 minutes per day. The initial planned mission duration was now 30 days, with a capability of extended missions to 50 days.

The technical evolution of the fourth generation reconnaissance camera, sponsored and funded by the CIA, involved the technical participation of both the Itek Corporation and the Perkin-Elmer Corporation. Beginning with the Itek study in the spring of 1964, to the final improved version of the "optical bar" configuration developed by Perkin-Elmer, required over two and one half years of engineering effort.

Perkin-Elmer submitted its proposal for the new camera on 21 July 1965.[11] However, this was not the end. After an initial evaluation, the NRO requested additional supporting data from the camera contractors and asked that technical personnel from these companies attend a three-hour briefing and be prepared to answer questions. The meetings were held on 2 and 3 August 1966.[12]

On 8 and 9 August, two proposal evaluation teams, both technical and operations, visited Perkin-Elmer.[13] In addition to a formal presentation by Perkin-Elmer management and program personnel, the task groups held working sessions with various production and logistics personnel and examined the facilities which were scheduled to be used on the new program.

On 1 September 1966, the Source Selection Board, headed by Chairman Leslie C. Dirks, reported their findings to Dr. Flax (Director of the NRO). The report suggested that certain deficiencies existed in the proposals and indicated the need for a careful reassessment of the impact on cost and schedule to rectify these deficiencies.[14]

Members of the Office of Special

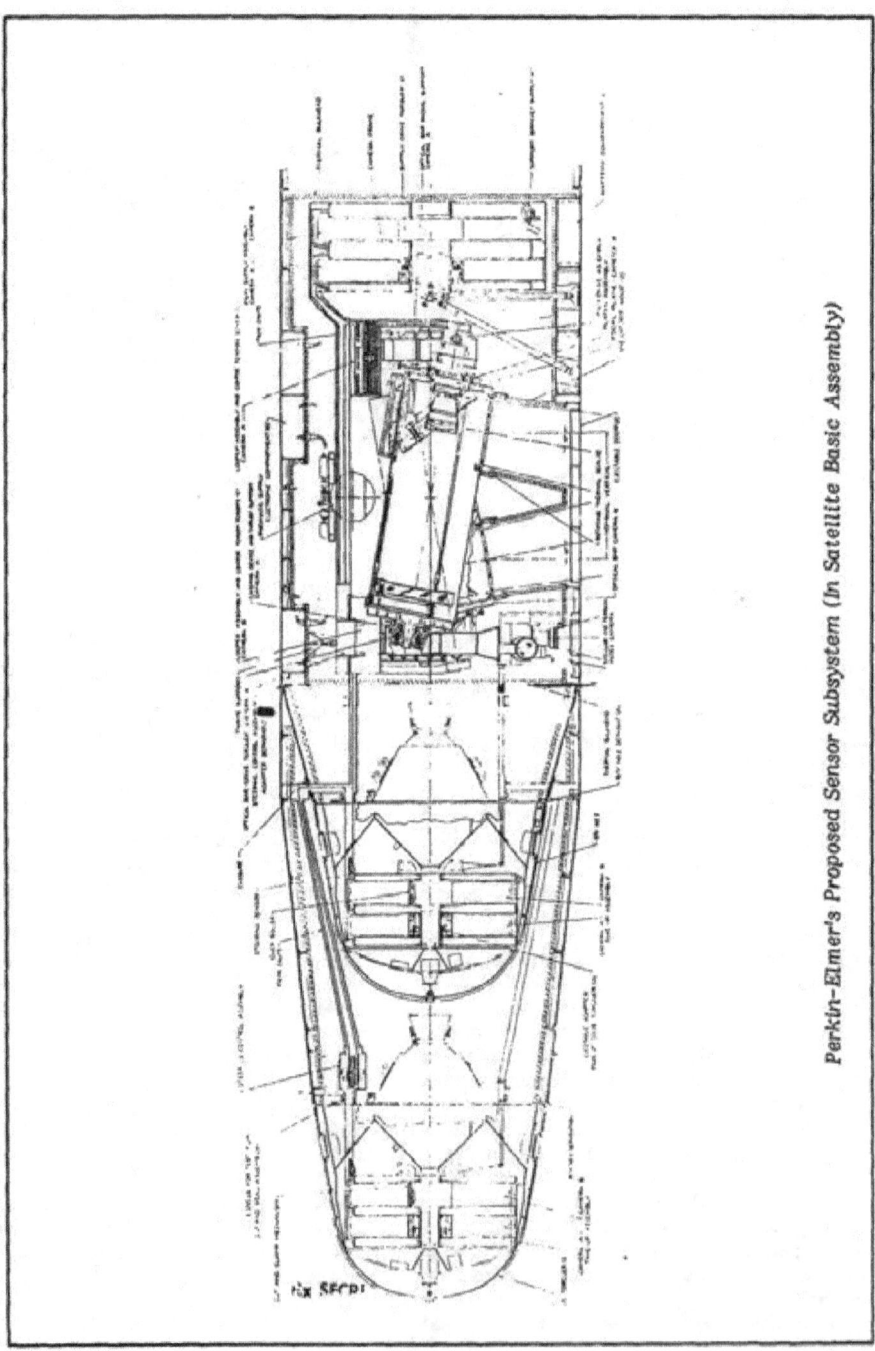

Projects/CIA visited Perkin-Elmer on 8 September 1966 to establish ground rules for the reassessment of the development schedules and associated costs. Perkin-Elmer's response was hand-carried to the Source Selection Board on 26 September 1966.

Program personnel at Perkin-Elmer continued to work on useful engineering analyses and experimental tasks to the day the winning proposal was announced by the NRO.

AWARD OF CONTRACT

On 10 October 1966, a phone rang in Bob Sorensen's office. The call was from John Crowley, Director of Special Projects in the CIA. Unfortunately, Bob Sorensen, who at that time was Senior Vice President of the Optical Goup, was attending a meeting outside of the plant. However, his secretary was finally able to reach him and inform him of the phone call. Upon contacting John Crowley, Bob Sorensen was given the good news. Perkin-Elmer had won the competition for the fourth generation reconnaissance camera, the Hexagon program. Two days later, the award of contract was confirmed officially by a TWX message that read in part, "This is to advise that your proposal has been reviewed and evaluated by the government and you have been selected for the award of a contract for essentially the effort as outlined in reference proposal."[1]

As soon as Bob Sorensen learned of the award, he called Dick Werner who headed the proposal activity and was to become the General Manager of the new Optical Technology Division designated to implement the Hexagon program. It was to be the third division in the Optical Group. The following morning, Dick Werner called all his program people together in a large engineering work area. He jumped up on a table, lit up the largest cigar any of them had ever seen and announced that Perkin-Elmer had won the competition. After a momentary sigh of relief, there was an immediate shout of joy that echoed through the halls. After the hand-shaking and back-slapping were over, Dick Werner said, "Now let's roll up our sleeves and get to work." And work they did! The waiting period was over and they could now move full speed ahead.

W. Richard Werner headed the Hexagon proposal team and later became the first General Manager of the Optical Technology Division.

The technical and administrative problems ahead, however, would prove to be ten times more difficult than anyone had envisioned. In many instances, this program was to change the careers of many in the company, and to some extent, have a personal impact on their lives. This was not to be a 9 to 5 job. It was a task that would test the temper and character of many assigned to the program. Some would become stronger because of their involvement, some would be hurt emotionally and physically, and some would fall by the wayside to be replaced by stronger ones.

One thing was clear at the outset — the success of this program could never be compromised because of any individual or situation. The program came first and accommodations were made by program personnel at all levels. If it meant working straight through a 24-hour day to solve a critical problem or finish a vital test, there was no hesitation on anyone's part. This attitude prevails to the present.

The weeks following the award were

very active. A kickoff meeting was held 19-20 October 1966.² It was a complete review of the Hexagon program organizations including the CIA, the System Engineering Contractor (CIA technical consultants), and the Sensor Subsystem Contractor (Perkin-Elmer). Hexagon program responsibilities were discussed and interim operating ground rules were established. Working group sessions included the Assembly and Checkout Group, the Security Group, and the Contracts Group.

A preliminary event list was compiled. Among the key milestones established were the letter contract award, the completion of the program plan, and the issuance of the formal contract award. To cover the period between 10 October 1966 and the day that the formal contract was to be issued (April 1967), negotiations were held with ▓▓▓▓▓▓▓▓ (CIA Contracting Officer) and Thomas Kindilien (OTD Director of Contract Administration and Purchasing). Perkin-Elmer was authorized to proceed with the Hexagon program in accordance with a statement of work outlined in a TWX message for a six-month period.³ A letter contract was eventually received by Perkin-Elmer on 23 November 1966.

The organization of the new Perkin-Elmer division was now in effect. Dick Werner, General Manager of OTD reported to Bob Sorensen, Senior Vice President of the Optical Group. On the day that the contract was awarded, OTD had a total of 217 people.⁴

To understand the status of the Hexagon program at OTD the day the contract was awarded, it is helpful to review the tasks that were worked on from the beginning of the program (23 June 1964, to 10 October 1966). During this period most program personnel were located in the Wilton facilities, both 50 Danbury Road and 77 Danbury Road.

The first statement of work covering Phase 0 (June 1964 to Sept 1964) specified that, "The attention of the study effort will be directed toward definition of the optimum form or forms of system, with particular emphasis on factors such as size, weight, power, and environmental requirements, compatibility with film recovery systems, control methods, operational reliability, and the delineation of special problem areas." The program was planned in three steps; a parametric study, investigation of alternate configuration concepts, and determination of the significant problem areas. The original intent was to complete all three steps and submit a final report on 28 September 1964. However, work did not proceed at the anticipated pace with the result that on that date, the program had progressed nearly through step two, with preliminary work accomplished in step three.⁵

At a meeting between CIA representatives and Perkin-Elmer on 14 September 1964, it was decided that the program be continued until 1 January 1965. An additional statement of work was agreed on and included the following:⁶

(1) Continuation of preliminary design effort on the three alternate systems to the point where a choice can reasonably be made on technical grounds.

(2) Completion of optical design investigation of transfer lens system.

(3) Investigation of automatic focus control feasibility.

(4) Feasibility and effectiveness of exposure control.

(5) Study of techniques for closed loop Image Motion Compensation.

(6) Determination of desirable characteristics of an optimum film.

(7) Analysis of interfaces and interactions of photographic system with vehicle in the case of all three systems (i.e., momentum, roll-joints, attitude control, etc.).

The delivery schedule of the final report on Phase 0 was changed to 15 October 1964. It was delayed and finally delivered on 16 November 1964.

During the time that the Phase 0 activity was in progress, the CIA also requested Perkin-Elmer's involvement in a detailed Phase I study and design effort for a production model of the fourth generation reconnaissance camera. The work statement agreed on by the CIA and Perkin-Elmer on 29 September 1964 included the following:[7]

(1) By 2 October 1964 (changed to 19 October)[8], select the single most desirable system for detailed Phase I study and design. Thereafter:
   (a) Proceed with the detailed optical design (study), complete Class "A" drawings of the optical elements, determine lead times and specify manufacturing tooling and test procedures.
   (b) Proceed with the design of a film transport subsystem. Breadboard elements of this subsystem for delivery and deliver an evaluation report with test results.
   (c) Proceed with the camera design to the point where drawings suitable for the engineering model can be prepared. Some mockups will be made.
   (d) Accomplish the necessary theoretical analyses to define the thermal, stress and other environmental parameters critical to system design and performance.
   (e) Complete detailed performance predictions.
   (f) Investigate and define weight, power, interfaces (thermal, mechanical, and electrical), coverage, and reliability.
(2) Carry out a design of reimaging optics.
(3) Design and breadboard an autofocus mechanism for delivery, and deliver an evaluation report with test results.
(4) Prepare a program plan to include:
   (a) Schedule for engineering model and prototype units.
   (b) Schedule of funding and other resources required to achieve a production run of 40 units at a rate of one per month, aimed at a first flight 1 November 1966.
   (c) Milestone schedule or PERT chart for critical items.
(5) Deliver a final report covering all the above.
(6) Submit monthly letter reports covering technical progress, manpower loading, and fund expenditure.

The final report on Phase I was delivered to the CIA on 1 February 1965. However, based on discussions in Washington, D.C., 26 January 1965, Perkin-Elmer was asked to do additional work on the following tasks during the month of February.[9]

(a) On existing ▓▓▓ samples, measure critical physical properties. Analyze optimum structure for molded ▓▓▓ and the new ▓▓▓ fused silica.
(b) Mockup full size figure 8 dynamic system for one cone to verify feasibility. Study tradeoff of duty cycle and accelerations.
(c) Complete analysis of film position when supported by two rollers in platen.
(d) Investigate and identify cause of high frequency film flutter in vacuum by experimentally varying hole patterns (in gas bars). Measure gas flow.
(e) Obtain further laboratory data on autofocus breadboard to test theory through wider range of conditions. Extend the breadboard to include Phase sensitive demodulation for signal processing.

(f) Project management reports on the above five tasks, travel, extension of leases on equipment for one month, guards for one month, and phones.

In the same message listing the above tasks, a Perkin-Elmer letter is referenced (MW-AH-32, 21 January 1965) discussing GEMS and AIM. These activities for the CIA began on 21 January 1965, were to be completed by 30 March 1965, and later extended to 7 June, 1965 and included two tasks.[10]

Task I Preparation of 12 GEMS. Each Gem shall be a simulated enlargement of a camera negative. The detailed general description and detailed specifications are specified in Attachment I to the subject contract.

Task II To measure the 3-bar target modulation detectability of Kodak films type 4404, type SO-206, and type SO-121 as a function of exposure, processing, and shape of modulation transfer function curve. The detailed requirements are specified in Attachment II of the subject contract.

On 1 March 1965, Chester Nimitz, Jr. received an important phone call. John Crowley of the CIA asked him to consider undertaking the design of the "optical bar" concept developed by the Itek Corporation. The CIA apparently had a stronger interest in the optical bar concept than it had in Perkin-Elmer's Phase I design, the "cocktail shaker." (Later Eastman Kodak's "M" system, also developed in Phase I of the Fulcrum program, would be continued by the Itek Corporation because of Kodak's involvement in the Manned Orbiting Laboratory program.) Chester Nimitz, Jr. replied that Perkin-Elmer would be unable to accept the assignment unless it had an opportunity to review and analyze the optical bar system. The CIA agreed to this arrangement and a statement of work covering this activity at Perkin-Elmer was approved. It included the following tasks:[11]

(a) Commencing 1 March 1965 continue useful work as outlined in Ref. 1 (TWX Message ▓ 3421) plus review of material (Itek design reports) furnished (Perkin-Elmer) at ▓ (CIA headquarters) on 5 March. This review to continue thru 22 March 1965.

(b) Commencing 22 March 1965 and continuing thru 20 April 1965 (changed to 7 May 1965)[12] complete the following tasks as indicated.
(1) Continue analysis of Fulcrum system as described by reports, drawings, breadboards, etc. furnished as GFE including appropriate experimentation.
(2) In conjunction with CIA technical personnel (consultants) and associate contractors, introduce new building blocks where useful.
(3) Layout facility and program plan for production program by May 1965.
(4) Continue buildup of manpower as indicated during 22 March meeting at Perkin-Elmer.

The above statement of work was later revised as follows:[13]

Following is revision to TWX 2450 referencing statement of work for letter contract based on TWX ▓ 7382 and 7383.

(a) Commencing 1 March 1965 continue pertinent tasks under work statement in (contract)

AM-7002 and review material furnished us (Perkin-Elmer) at ▊ on 5 March 1965. This review to continue through 22 March 1965.

(b) Commencing 22 March 1965 and continuing thru 20 May 1965 complete the following tasks as indicated.
   (1) Continue analysis of Fulcrum system as described by reports, drawings, breadboards, etc., furnished as GFE including appropriate experimentation.
   (2) In conjunction with your technical personnel and associate contractors, introduce new building blocks where useful.
   (3) In addition to above furnish preliminary design layout for second configuration.
   (4) Furnish a program plan by 20 May 1965 covering the following: (a) the technical content and justification for the system approach, (b) the schedule and ROM costs for the system proposed along with carefully stated assumptions, and (c) a complete presentation of resources applicable, i.e., facilities and manpower, both corporate and subcontract and the proposed management arrangement internal to your company.
   (5) Commencing 21 May 1965 the contractor shall continue design effort and brassboarding and coordination with associate contractors leading to interface definition to the extent possible by 1 June 1965.

One of the important additions in the revised statement of work was the inclusion of a second configuration ("M" system). This system was proposed by Dr. Kenneth Macleish (Vice President, Engineering, Optical Group) who was concerned about the complexity and cost of the Fulcrum design. He felt that it was important to show the CIA that there was an alternate solution. The "F" design was predicated on being able to see 120° in one sweep and as a result it was poorly adapted to seeing small angular sectors. He was proved correct in later years since it was troublesome to program in small "looks" efficiently.[14]

During the Phase I effort, Perkin-Elmer also negotiated a contract with the CIA to perform the following tasks between 15 January 1965 and 30 November 1966.[15]

Task I - Continuous Polishing

(1) Continued experimentation of polishing process with breadboard and leased machines.
(2) Continued development of lap interferometer.
(3) Completion of environmentally controlled polishing room.
(4) Evaluation of the process, due 15 May 1966.
(5) Final written report (48" lapmaster) due 15 May 1966.

Task II - Selective Coating

(1) Continued development of the coating techniques with an objective of successfully correcting moderate size mirrors during this phase.
(2) Combined development of suitable coating materials and coating parameters.
(3) Evaluation and written report due on 15 May 1966.

Task III - Optical Test Techniques

(1) Develop Hologram interferometer and prove technique which will show deviation of test piece from master. Application

to selective coating deposition; aspheric manufacturing and continuous polishing action will be determined.

(2) Report due on 15 May 1966.

Task IV - Image Quality

(1) Conduct studies and tests in the continuation of the search for a summary measure of image quality in accordance with contractor's proposal dated 2 August 1965. Said proposal being incorporated herein by reference. Reports shall be submitted on a monthly basis and final reports submitted on the completion of each task. All work shall be completed 30 November 1966.

Task V - Herriott Interferometer

(Amendment 1 - started 1 April 1966.)

Task VI - Fizeau Interferometer Including Skip Interferometer

(Amendment 2 - started 15 January 1966.)

By the beginning of June 1965, Perkin-Elmer was preparing for a presentation which was to be given to the Land Panel, a subcommittee of the NRO formed to study various proposed systems for the fourth generation reconnaissance system, to decide the direction of the program. A statement of work which covered the activities for the month of June 1965 contained the following tasks.[16]

Task 1 Conduct the maximum amount of interfacing with the associated contractors such as to have the best possible integrated F-prime system by 30 June. (F-prime was the Perkin-Elmer redesign of Itek's "optical bar" Fulcrum system). Redesign the current F-Prime configuration only to the extent that these interfaces dictate.

Task 2 Conduct sufficient interfacing with the associate contractors such as to have a valid integrated Matchbox system by 30 June. (The "M" system was the configuration recommended by Dr. Macleish). Redesign the current matchbox configuration only to the extent that these interfaces dictate.

Task 3 Perform sufficient analysis to support the Matchbox performance predictions so that a meaningful comparison can be made with the F-Prime performance predictions.

Task 4 Design and construct a breadboard of the 180 degree twister as discussed with ▆▆▆ on 3 June. Subject breadboard will be completed and results deliverable to the panel by 30 June. (The Land Panel meeting was eventually held on 21 July 1965).

Fulcrum program activities at Perkin-Elmer for the month of July 1965 were covered by a work statement which included the following tasks.[17]

Task 1 Complete construction and debug of breadboard of 180° twister by 7 July. Submit test plan to Headquarters by 7 July. Complete tests by 16 July. Submit report on test results by 20 July.

Task 2 Design the Matchbox

	Camera System such as to include those electronic features best suited for total system performance (i.e., servomechanisms for driving the oscillating mirror and/or IMC motions including earth rotation). A dual approach (mechanical and electronic) will be maintained until such a time as the Project Office and Perkin-Elmer can make a clear choice.	Task 6	Provide to the reentry vehicle contractor revised weight estimates of payload equipment to be housed within the reentry vehicle. Conduct sufficient experiments to ascertain for the reentry vehicle contractor the amount of c.g. offset from the spool axis due to operational film takeup conditions.
Task 3	Conduct a complete thermal analysis of the Matchbox Camera System for worst case conditions during maximum photographic sequencing and during "stored" period. Report of results is due 20 July.	Task 7	Institute an on-going weight, balance, and power distribution reporting system to provide such information to the SEAC contractor.
		Task 8	Perform a revised reliability study based on the reliability bogey provided by the SEAC contractor.
Task 4	Formulate dynamic models for the F-Prime and Matchbox Camera Systems. Identify the disturbance pulse shapes and perform a dynamic response analysis. Generate optical bar tolerance criteria in the case of F-Prime showing allowable slopes and deflections at critical locations. A preliminary version of each model is due by 9 July.	Task 9	Size, in detail, the structural system for both the F-Prime and the Matchbox Camera Systems.
		Task 10	Begin System Specifications Books for both the F-Prime and the Matchbox Camera Systems.
		Task 11	Investigate in detail the synchronization of the optical bar of the F-Prime System to the oscillating platen and prepare a report on the technique selected with assigned tolerances by 16 July.
Task 5	Prepare a three-sigma systematic and random image blur summary in both the scan and forward directions with the equivalent error budget. Prepare a three-sigma focus error budget summary. These summaries are required for both camera systems by 20 July.	Task 12	Determine the central aperture obscuration for the F-Prime optical system. Describe in detail why the 13% figure determined by Itek cannot be met, if this is true.

On 31 July 1965, the CIA sent a

message to Perkin-Elmer confirming telecons to the effect that the Fulcrum Program would embark on a Program Definition and Final Configuration Study Phase.[18] The following work statement defines the tasks to be accomplished in a three-month period.[19]

This statement of work defines the work to be accomplished by the Perkin-Elmer Corporation during August, September and October. The contractor shall provide all necessary manpower and material to support the tasks defined herein in accordance with the attached schedule. During this period, the primary aim of all tasks shall be the presentation of a complete accurate design report on or about 22 November. The following configurations are defined for this period.

    Configuration A - Titan II, 60" F.L. camera, 15 day mission, 880 pounds of film, one R/V.

    Configuration B - Titan SX/SRM, approximate 60" F.L. camera, 30-day mission, 3100 pounds of film, two or more R/V's.

    Configuration C - Titan SX/SRM, 60" F.L. camera, 45-day mission, 2500 pounds of film, two or more R/V's.

Task 1   Prepare a briefing to be given 14 September 1965 for Headquarters comparing the F' and M' systems with recommendation of system for further study as configuration "A". A summary report documenting this briefing shall be submitted by 1 October.

Task 2   Define the selected "A" configuration payload design and present a design review for customer approval on 15 October. A design review package shall be prepared and submitted to the customer by 11 October. This design review package shall contain but not be limited to system description, layout drawings, system block diagram, parts count and system reliability estimate, performance calculations, and mass properties and power estimates.

Task 3   Prepare a program plan for the acquisition and operational (3 years) phases of Configuration "A" to be submitted on 1 October. (This means 3 years supply of flight articles, i.e., one per month or 36 units). This program plan shall include but not be limited to detailed schedules, cost, development test plan, and identification of long-lead items with schedules and cost by month for the first six months, including facilities and hardware.

Task 4   Conduct a conceptual design and performance estimate for an enlarged payload (Configuration "B") based on the selected 60" F.L. version. Conceptual design to emphasize the optical system design and evaluation of film handling problems. Conduct a briefing on this design on 15 October. Documentation of this briefing to be submitted by 29 October.

Task 5   Conduct a design review of the selected large payload (extended service) configuration "B" or "C" for the customer on 17 November. A design review package as de-

Task 6 — Prepare a program plan for the acquisition and operational (3 years) phases of the large payload (as defined in Task 4) to be submitted on 17 November.

Task 7 — Participate in a formal briefing on or about 22 November 1965 and other briefings as necessary.

Task 8 — Support SEAC and other associate contractors as required in: (a) Factory to launch systems requirements analysis for Configuration "A" (This means the flow from factory to launch); (b) developing overall system program plans; (c) overall system planning; (d) reliability allocation and assessment; (e) definition of interface requirements; and (f) maintenance of liaison with other agencies.

Task 9 — Maintain weight, balance, and power distribution reporting on all configurations.

Task 10 — Conduct reliability studies on all configurations based on reliability bogeys provided by the SEAC contractor.

Task 11 — Continue twister experiments to the degree necessary to demonstrate design feasibility.

Task 12 — Initiate experiments to develop selected designs in critical areas.

Task 13 — Study feasibility of use of autofocus as applicable to the selected configurations.

Task 14 — Submit informal weekly TWX progress reports starting 20 August 1965.

Task 15 — Prepare and submit a final report summarizing the work performed under the contract.

Shortly after the 14 September 1965 briefing, given to the CIA by Perkin-Elmer in which Perkin-Elmer selected the F-Prime system, a revised work statement was written and included the following tasks.[20] Tasks 11 and 12 were later added to the work statement.

Task 1 — Prepare a briefing to be given 14 September 1965 for Headquarters comparing the F' and M' systems with recommendation of system for further study as Configuration "A" (Titan II - 60-inch F.L. camera - 15-day mission - 880 pounds of film - one R/V.). A summary report documenting this briefing shall be submitted by 1 October 1965.

Task 2 — Prepare proposal on selected configuration and present a design review for customer approval on 15 October. A design package shall be submitted to the customer by 11 October.

Task 3 — Support customer consultants and associate contractor's as required to develop overall program plans and interfaces.

Task 4 — Conduct reliability stud-

Task 5 — Initiate experiments to support selected designs in critical areas.

Task 6 — Study feasibility of use of autofocus as applicable to the selected configuration.

Task 7 — Prepare System Specification Book incorporating previously uncollated material which will be supplemented by additional specification material as produced. Preliminary issue AH65-1165. Submitted 1 Nov 1965.

Task 8 — Additional project office tasks.

Task 9 — Submit informal biweekly TWX report.

Task 10 — Prepare and submit a final report summarizing work performed under this contract.

Task 11 — 3404 Evaluation (new task) ▮▮ 2900 18 Nov 1965 (report submitted ▮▮ 2933 2 Dec 1965.)

Task 12 — (new task ▮▮ 2900 18 Nov 1965.) Image quality. Gems

Amendment No. 2 on Contract FS-2057 covered the period of performance from 1 August to 31 December 1965 and included experiments which were to be continued and initiated.[21,22] System experiments included the following: film flatness experiments, film transport properties, film handling in vacuum, film outgassing, and autofocus experiments. Optical experiments included material stability, Herriot interferometer development, Fizeau glass investigations, and continuous polishing, selective coating and optical test techniques. Support activities included system test planning, reliability studies, and quality control planning.

On 13 January 1966, Perkin-Elmer submitted a program plan to the CIA which incorporated all of the current and planned program tasks and activities at Perkin-Elmer.[23] This program plan was the basis of all the effort at Perkin-Elmer through the following months. Effort was continued through the proposal preparation (which began on 23 May 1966 when Perkin-Elmer received the RFP, to the day the proposal was submitted to the government on 21 July 1966) to the award of contract on 10 October 1966.

COVER AND SECURITY CONSIDERATIONS

Perkin-Elmer's involvement in classified programs began even before the reorganization of its activities into commercial and government business. Prior to divisionalization, in the spring of 1956, the responsibility for the security aspects of classified programs resided primarily in project management and administration. With the exception of a program codenamed Projector Project (optical instruments for the U-2 aircraft), most of the programs at Perkin-Elmer at that time were either unclassified or classified at lower levels (confidential or restricted) and did not require special secure facilities. Perkin-Elmer received its secret facilities clearance on 13 March 1956.

Shortly after Perkin-Elmer received its secret facilities clearance, the company hired Patrick Murphy to oversee the security requirements of all classified programs at Perkin-Elmer. After Murphy left the company, James McNamara assumed this role. After his resignation, Herbert Dunning, who was hired in 1965, became Chief Security Officer. However, it was not until Robert Markin joined the company in June 1966 that a full-time security administrator was assigned to the

Hexagon program. Markin was Dunning's assistant for three weeks and then became the Security Chief of the Hexagon program. He had worked at the Central Intelligence Agency (CIA) prior to joining Perkin-Elmer. After Perkin-Elmer was awarded the Hexagon program, Markin became the OTD Security Chief.

At the beginning of the space reconnaissance program (Discovery missions), there was no formal security control system. Although there were security personnel at the CIA who managed this activity and provided security accommodations for "black" programs, a formal industrial security manual for national policy guidance of space reconnaissance programs did not exist. Classified contracts awarded to industry and research laboratories contained only general instructions and guidelines on managing program security. The details for providing a secure area and developing policies to protect classified program information was the responsibility of the company awarded the contract.

The U.S. Air Force began to develop the "Byeman" Industrial Security Control System Manual for overhead space reconnaissance programs in the early 1960's. CIA established security policy was implemented through the Byeman Manual. In addition both the CIA and the U.S. Air Force adopted the procedures issued by the U.S. Intelligence Board on physical security construction criteria. It should be noted that quite a few of the instructions during the initial stages of the overhead reconnaissance programs to the industrial contractors were through "word of mouth."

During the first few years that the Byeman manual was being developed, a number of government intelligence and industry security representatives, including Markin, participated in meetings to discuss the various aspects of protecting "black" programs. Over a dozen meetings were held, both on the East Coast and the West Coast, to create a practical and functioning document. The Byeman Manual is now the basic security document used on all programs related to covert overhead reconnaissance.

In 1966, when the Hexagon program was awarded to Perkin-Elmer, company net sales were $88,000,000. The Hexagon program had an initial contract value of ████████. When Perkin-Elmer was instructed to "make the money disappear" and not let anyone outside of the program know that a massive contract had been awarded to Perkin-Elmer, it faced a difficult problem. For a small $88 million company to try to hide ████████ worth of activity was a task equivalent to trying to hide an elephant in a closet. And the question became, "How do you hide an elephant?" If Perkin-Elmer had been a large company doing two to three billion dollars worth of business annually, it would be a relatively simple task to shield the existence of the Hexagon program. Perkin-Elmer's solution to the problem of hiding a massive covert program was to deny its existence and answering the question with another question, "What elephant?"

At the time (January 1965) that the Phase I proposal was being written by Perkin-Elmer, only general security instructions were provided by the agency. A section of the Perkin-Elmer proposal which was submitted to the agency in January, 1965, discussed, in general terms, how security on the program would be handled if the contract was awarded to Perkin-Elmer.[1] The introductory paragraph in the report stated, "Security measures to be followed on this program will, in general, follow the patterns established on previous covert programs. The work will be done, for the largest part, in a separate secure area and every effort will be made to keep unauthorized persons from learning what is being done, what it is to be used for, its capabilities, the identity of the customer or associated contractors, or the final schedule of operations." It more or less expressed the security policies followed on the Fulcrum program from June 1964 to the award of the Hexagon contract in October 1966.

The security writeup in the Phase I proposal also addressed the problem of concealing the program from the public and uncleared Perkin-Elmer employees. "The size of the program will make it impossible to hide the existence of the program within the company. Thus, a cover story must be derived. Any cover story comprehensive enough to explain all the items requiring explanation (number and type of personnel, nature of materials ordered, etc.) must describe specific hardware to be convincing. Yet, any specific cover story is subject to easy refutation since all firms already working on similar sytems would be aware that the story is false, and all the military sharing a legitimate interest in whatever type of specific hardware is described would claim a legitimate need-to-know about the program. The proposed basic story theme, therefore, is that a number of projects are being worked on, some to be classified, some Company Private, and some open. Instead of one massive program, there would be a number of smaller ones, requiring less explanation. They would be grouped, more or less, together with an explanation that there is an overlapping of personnel on the various projects. Separate internal work orders would be written to cover the various programs. This procedure is similar to that used on other covert programs and fits well with the pattern of work, accounting, and control within the company." This section of the report also discussed the method of handling communications, guard personnel, alarm systems, and document and visitor control systems. It was a carefully considered security plan.

The CIA's Request For Proposal (RFP) for the Hexagon Program sent to Perkin-Elmer on 23 May 1966 contained only general instructions on handling a classified program.[2] One of the requirements of the RFP was a company plan for a security program. The plan submitted by Perkin-Elmer was almost a verbatim copy of the security approach contained in the Fulcrum (January 1965) proposal, with a few modifications.

After the Hexagon program was awarded to Perkin-Elmer in October 1966, the CIA requested the OTD Security Office to design a detailed security plan for the program. Markin responded by writing a General Security Bill, Guard Orders, Classification of Project documents, and a Security Classification Guide and a Courier Proposal.[3,4,5,6,7] During discussions with Agency representatives responsible for overall program security, several basic guidelines were established. Perkin-Elmer could not reveal that it had any association with satellite surveillance or that a camera was being designed for the CIA. It was realized by all involved that the technical problems of designing one of the most sophisticated and complicated cameras ever envisioned would be extremely difficult. However, solving these technical problems, and at the same time meeting a tight schedule and maintaining complete secrecy magnified the difficulties tremendously.

Perkin-Elmer used an in-house contract numbering system called the Sales Purchase Order (SPO) system. The Electro-Optical Division listing at that time contained over 50 SPO's and about 20 internal Work Orders (WO's). It was Perkin-Elmer's plan to establish a separate division for the Hexagon program using a similar SPO and WO listing. After OTD's creation in October 1966, the OTD Contracting Office reviewed the entire Hexagon program and established almost 20 separate tasks which were required to design, develop and fabricate the six sensor subsystems that Perkin-Elmer was initially contracted to deliver to the CIA.

This system encompassed a large variety of tasks such as the construction of the Danbury facility, thermal design, interface activity with associate contractors, and the design and fabrication of the major camera components and assemblies. Perkin-Elmer created a separate contract order for each of these tasks, and then using the theory of "plausible denial" created a cover story and made the new Optical Technology Division appear similar to the other Perkin-Elmer divisions. By "plausibly denying" the true nature of the camera components and assemblies that were being built or purchased, in total or in part — simply by calling them something else — normal "white" procurement procedures could be used. This theory was used throughout the program activities to minimize the need for time-consuming and expensive covert procedures. While it was important to keep inviolate the security of the program, the main objective was to build and fly workable hardware, within schedule, and at a minimum cost.

At the time the Hexagon program was awarded to Perkin-Elmer, the newly established Optical Technology Division was understaffed and did not have an adequate number of people with the appropriate disciplines to assign to the program. Staffing a growing division is a difficult task under normal conditions, especially in a strong economic climate. Providing a work force for a very secretive technical program created major problems. One fact made known to people briefed by the OTD Security Officer was that Agency approval for their "access" to the program was not a "clearance."

Not only did the recruitment program dilute the efforts of accessed administrative and technical people already assigned to the program, but the people involved in recruiting new employees for the program were doing the job with "one hand tied behind their backs." The major question was, "What do we tell these guys about their assignments if we can't divulge the nature of the program?"

The recruitment advertisements in newspapers and trade magazines had to be sufficiently accurate to attract the required disciplines, and at the same time conceal the true purpose of the program. The first step in the cover story was that the new division was recruiting people not for a single major program, but for a multiplicity of contracts. The potential employees were told that they would be assigned to a number of programs, some of which were classified and some unclassified.

The new hires were asked to fill out various employee questionnaires, including a security clearance form. However, they could not be told that their employment was based on the successful completion of a background security check or that they would be placed in a "holding area" until an access was granted by the agency. These requirements created many problems. The mere fact that all hires would be isolated in one area for periods up to two to three months, and in some instances as much as ten months, and that as time passed by they would see their co-workers leave the holding area one by one, produced a compromising situation. The remaining unaccessed employees would soon realize that they needed an access before they transferred to a permanent assignment.

Perkin-Elmer was placed in a dilemma. The CIA's reply to this problem was, "They may speculate about their situation, but they cannot be told that approval is required for their transfer out of the holding area."[8] Perkin-Elmer could not tell the new hires that their employment was contingent upon an access approval and that if it was not granted, they would either be transferred out of OTD or terminated. Apparently, the CIA did not want a rejected employee to appeal the decision and place the Agency in a difficult position.

One additional problem that Perkin-Elmer faced if a new employee hired for the program was not granted access to the Hexagon program was that the CIA did not inform the company why the person was rejected. It was clearly none of Perkin-Elmer's business. The fact that certain people were not suitable for a special program access did not necessarily indicate that they were unsuitable for (secret or top secret) Department of Defense contracts.

The new hire holding area was located at 50 Danbury Road, Wilton, CT in a high bay area and became known as the "tank." An accessed engineering supervisor was assigned to oversee the work in this area. Work directed to this area consisted of program tasks that could be "sanitized." Using "plausible denial," various engineering and administrative assignments were given to people waiting for program access. Thereby some productive work could be accomplished. However, because of security constraints and because so many aspects of the program could not be "sterilized," the efficiency of this work group was never more than 50 to 60 percent.

Enormous pressures began to build on the program for additional accessed people. Under normal conditions, a security check required two to three months. However, because of the large number of

High Bay Area at 50 Danbury Road, Wilton, CT in 1965-66

security checks that were being processed by the CIA, and because the 1960's saw the passage of much social legislation which restricted the CIA from using commercial credit bureaus and other government agencies to collect personnel data, the process was slowed down considerably. The original staff of 12 people of Phase O of the Fulcrum project increased to a group of 50 people at the end of Phase I (January 1965). In the early part of the Program (February 1965), it required 2.6 months to acquire security access for a new hire. By October 1965, the waiting period increased to 4.5 months.9

While everyone on the program required access approval and was subjected to the same intensive investigation, certain facts were withheld even among the accessed personnel. After receiving program access from the CIA, the empoyees were briefed individually by OTD security officers.

There were three levels of briefing. Phase I, the lowest level, was given to people involved in the support functions of the program, such as the maintenance personnel and custodians. They were told that everything they heard or saw in the secure areas was considered classified and not to disclose what they learned to anyone — essentially, they were told nothing. A Phase II briefing was given to 95% of the remaining personnel. They, of course, needed to know the performance specifications, the purpose of the equipment, and the identity of the associated contractors. But Phase II personnel were told that the customer was the USAF. This was at the CIA's direction. A Phase III briefing was given mostly to program management and those with a need-to-know. They were told everything, including the true identity of the customer, the CIA. In addition, they were made aware of the different levels of briefing and were cautioned not to divulge this information to anyone, including Phase I and Phase II personnel. However, even Phase III program personnel were not given information indiscriminately unless they had a need-to-know (e.g. mission plans for each launch, areas photographed, etc.) and only people in a Perkin-Elmer group called the Post Flight Area (PFA) saw the actual photographs taken during each mission.

The security badges of Phase III personnel were coded to enable Phase III people to identify each other. Their badge numbers contained the digit "0". This badge coding technique could also be used to identify program personnel with other special need-to-know requirements (i.e., security officers, PFA personnel).

BUILDING PROGRAM

The Perkin-Elmer Hexagon proposal included a plan for providing engineering and administrative offices, manufacturing and test facilities, and assembly areas for the Hexagon program. How these building plans evolved is based in part on circumstances surrounding the beginning of the program in June 1964.

When Perkin-Elmer first became involved in the Ad Hoc program, personnel assigned to it were either located at the Connecticut Avenue facility in Norwalk, or the building at 50 Danbury Road, Wilton, Connecticut. By July 1964, everyone on the program was moved into a secure group of offices on the second floor of the Wilton building. As the group expanded, laboratory and storage space was made available in adjacent offices on the second floor.

The Phase I Ad Hoc Study Report, submitted to the CIA at the end of January 1965, indicated that the Perkin-Elmer facilities were adequate to initiate the program and could be expanded on available property in time to accommodate the total program.[1] The report stated that of the total project requirements of 96,000 square feet, approximately 50 percent would be used for subassembly, final assembly, and testing of the deliverable item. Of the 48,000 square feet, 24,500 square feet was standard instrument assembly space and could be provided by Perkin-Elmer's present facilities. The remaining 23,500 square feet would include 9,500 square feet of clean, high ceiling area with traveling crane equipment, and 14,000 square feet for environmental test equipment.[2]

Soon after the study report was submitted, Pete Clonan, Director of Corporate Facilities, supported by Ad Hoc administrative personnel, started looking for additional facilities in Fairfield County, Connecticut. After studying several available properties, it was decided that they were unsuitable since a substantial amount of money would be required to convert them for Ad Hoc program use.[3]

In the summer of 1965, property diagonally opposite the Wilton building became available. The building was originally constructed by the Hallicrafter Company and later taken over by the Manson Company. In September 1965, Perkin-Elmer purchased the 83,000 square foot plant on 22 acres at 77 Danbury Road, Wilton, specifically for the Ad Hoc program. The "Manson" building, as it was referred to at that time, was partially occupied by Ad Hoc personnel on November 1965, and by the final closing in 1 April 1966, most of the project personnel were moved to the new facility.[4]

The Perkin-Elmer Hexagon proposal submitted to the government on 21 July 1966 contained detailed facility plans for the program.[5] The "Manson" property offered convenient proximity to the work force residence, close relationship to Corporate and other supporting functions, and a suitable layout for initiating program tasks with only minor arrangements.

A two-phase plan would involve rearranging approximately half of this plant during the first six months of the program for optical fabrication and manufacturing fabrication and assembly. During rearrangements, these two functions would be carried out in other corporate facilities. Simultaneously, construction would start on a 110,000 square foot addition to the plant at 77 Danbury Road to accommodate all design, engineering, fabrication, assembly, and additional facilities for test activities. The second phase of the plan involved occupying these new facilities, starting at the end of the 12th month after award of contract.

Of importance to the program during the first phase was the availability of approximately 10,000 square feet of Class

"Manson" Building, 77 Danbury Road, Wilton, CT as it looks today

100,000 clean area in a new north wing of the Wilton facility beginning with program startup. This clean area was to be available until completion of the 110,000 square foot addition at the "Manson" site. At the completion of the two-phase plan, the entire program was to be housed in the expanded single plant at 77 Danbury Road, Wilton.[6]

An architectural and engineering study of the Environmental Test Facility was completed by Jackson and Moreland, a Boston company that also worked on Itek facility plans just prior to the CIA termination of the Itek Contract in the beginning of 1965.

Soon after Perkin-Elmer submitted the Hexagon proposal, it became apparent that due to the Town of Wilton zoning restrictions and the high cost ($1,000,000) of removing a small rock "mountain" on the Manson site, it would be less expensive to build a completely new facility at another site. Fortunately, the company had purchased some land (55 acres) at Wooster Heights in Danbury, Connecticut in the 1960's. It was decided to use this property for a facility designed specifically for the Hexagon program.

Studies to use this land for corporate expansion were in progress even before the Perkin-Elmer Hexagon proposal was submitted to the government in July 1966. The company had invested some risk funds to study the Wooster Heights site.[7] Architect and engineer consultants were preparing facility layouts for steel and foundation designs and by 11 August, test soil borings were in progress.[8] By the end of September, 1966, the architect and engineer consultants submitted steel drawings and a facility construction and actuation schedule to Perkin-Elmer.[9]

With the announcement on 10 October 1966 that Perkin-Elmer had won the competition, plans for the new Danbury facility moved ahead at a faster pace. Site preparation was started on 18 October. Construction of the new building moved along smoothly until 15 May 1967 when a Teamster Trade Union strike halted work.[10] The strike lasted until 28 June 1967 resulting in a 7½ week schedule slippage. Construction again moved ahead, and by 9 February 1968, the second floor of Building 1 (the facility consisted of five

The Manson Building "Mountain"

connecting wings) was partially occupied by Perkin-Elmer personnel.[11,12]

A major move from 77 Danbury Road, Wilton occurred on 29 March 1968 when several Optical Technology Division departments were assigned to the Danbury facility. A brochure, distributed to project personnel, explained the arrangement of the new facility, planned working hours, etc.[13] A memo from Ken Patrick, General Manager of the Optical Technology Division, described a company procedure that would reimburse personnel for any additional travel (for a period of six months) incurred because of the move to the new facility.[14]

Large Chambers Under Construction

Wooster Heights Facility Construction

Work continued on the unfinished wings of the facility and by the beginning of 1969, construction was completed. By that time, most of the program personnel were in the new location or in one of several buildings nearby the Danbury facility. The new Danbury facility contained over 270,000 square feet, with additional storage provided by two Butler buildings on the site. This was almost three times more space than estimated in the January 1965 Ad Hoc proposal.

Chamber A Final Assembly

During one of the facility meetings, there was some disagreement on the manner in which the exterior of the Danbury building was to be finished. The haggling continued until Chester W. Nimitz, Jr., the President of Perkin-Elmer, joined the meeting. The debate continued about the enormous costs that were being stacked up for the building. Nimitz, with feet resting on the table said, "Tell me how much it's costing to put the brick veneer on the building to make it look like the Corporate building in Norwalk." Someone replied, "$250,000." Nimitz shot back, "Leave the damn bricks off."[15] Five words, each worth $50,000.

Completed Danbury Facility

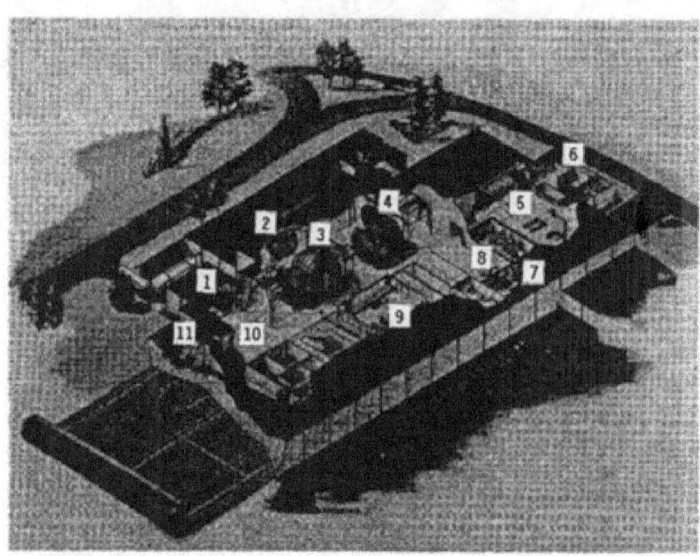

Overall Schematic of Danbury Facility

1. CHAMBER "D"
2. VACUUM PUMPING FACILITY
3. CHAMBER "A"
4. CHAMBER "B"
5. VIBRATION ENVIRONMENT FACILITY
6. ACOUSTIC ENVIRONMENT FACILITY
7. 6' X 6' VACUUM CHAMBER
8. 10' X 12' THERMAL VACUUM CHAMBER
9. CLIMATICS CONTROL ROOM - SSTC
10. FINAL ASSEMBLY AREA
11. AIRLOCK

SENSOR SUBSYSTEM DESCRIPTION

This history covers Hexagon program activities at Perkin-Elmer to the end of 1983 and includes the launch of satellite vehicle No. 18 (SV-18). The following is a description of the Sensor Subsystem for Hexagon program satellite vehicles SV-17 through SV-20.[1]

The performance requirements of the Hexagon program evolved from specifications used in initial studies by Perkin-Elmer in the spring of 1964. The original exhibit described the requirements for a space reconnaissance system carrying 880 pounds of film (EK Type 4404 - 7-inch wide, 68,000 feet) for a 5-10 day mission to be recovered in a single recovery vehicle.[1] By the time the first Sensor Subsystem was ready for launch in June 1971, the satellite vehicle design was changed to carry four recovery vehicles and the film payload was increased to 1,576 pounds (EK 1414 UTB, 6.6-inch wide, 208,000 feet). SV-1 flew a 31-day mission and transported over 175,601 feet of film. SV-17, launched in May 1982, flew a 261-day mission and transported 303,527 feet of film (Type SO-315/SO-130)[2].

The Hexagon program's satellite vehicles are orbital photographic reconnaissance systems with search, surveillance, and mapping capabilities. Each SV, launched by an aerospace vehicle with its Titan III D Booster, contains within its three sections a Sensor Subsystem (SS), various orbit, tracking, telemetry, and other control systems, a Solid State Sensor Camera System, and four SS film payload recovery vehicles.

Launched and injected into earth orbit, the satellite vehicle is commanded to operate and control the Sensor Subsytem throughout the photographic mission. After each recovery vehicle accumulates exposed photographic film from the Sensor Subsystem (a stereo camera assembly), it is separated from the satellite vehicle and is air-recovered. After recovery, each film load is shipped to the film processing facility for processing and duplication, and then to various users for evaluation.

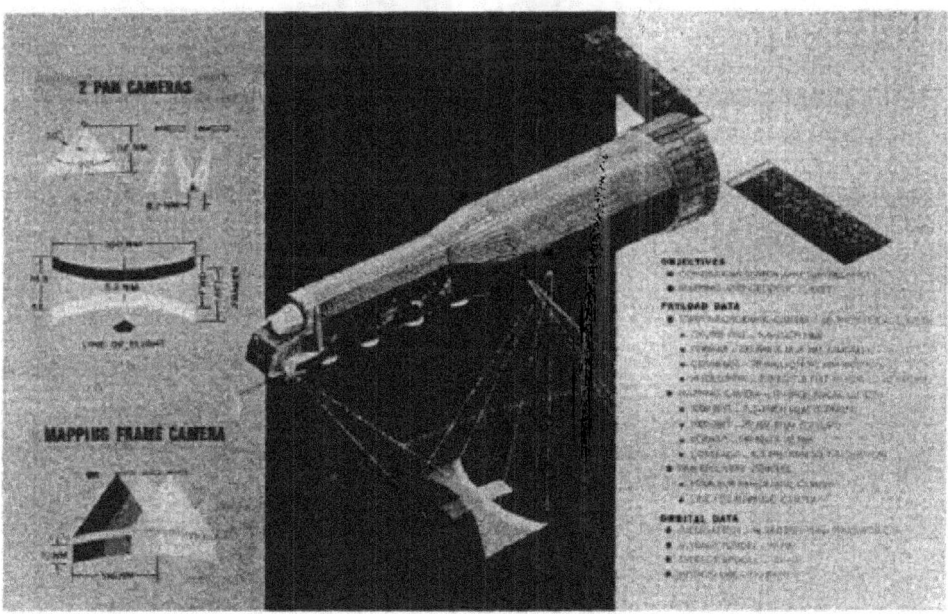

Hexagon Payload Operations (Typical of SV-5 through SV-16 Configuration)

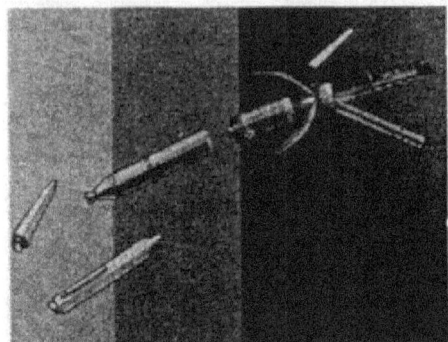
Booster and Shroud Separation

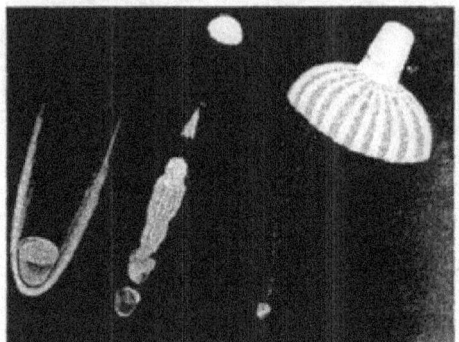
RV Atmospheric Re-Entry

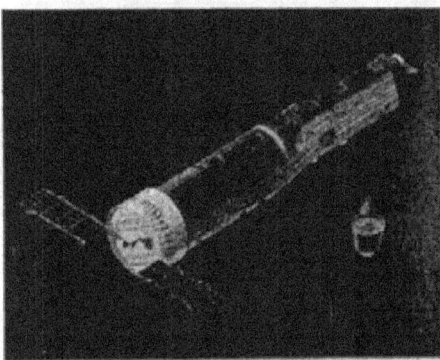
Recovery Vehicle RV Separation

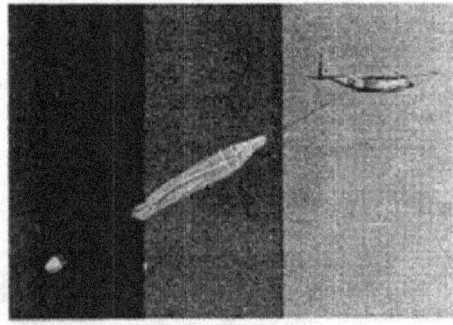
Mid-Air Capture of RV

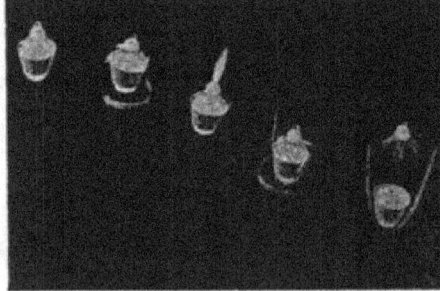
Recovery Vehicle Spin-Up

The three sections of the satellite vehicle (forward, mid, and aft) contain the following equipment, modules, and subsystems.

Forward Section

SS Recovery Vehicles (RV-1 through RV-4)

Mid Section

Two-Camera Assembly
Film Supply
Pneumatics and electrical power

Solid State Sensor (S^3) Camera System

Aft Section

Orbital Adjust System
Reaction Control System
Attitude Control System
Back-up Attitude Control
Electrical Distribution and Power
Telemetry, Tracking and Command
Supplementary Pneumatics Supply Module

Hexagon Camera System

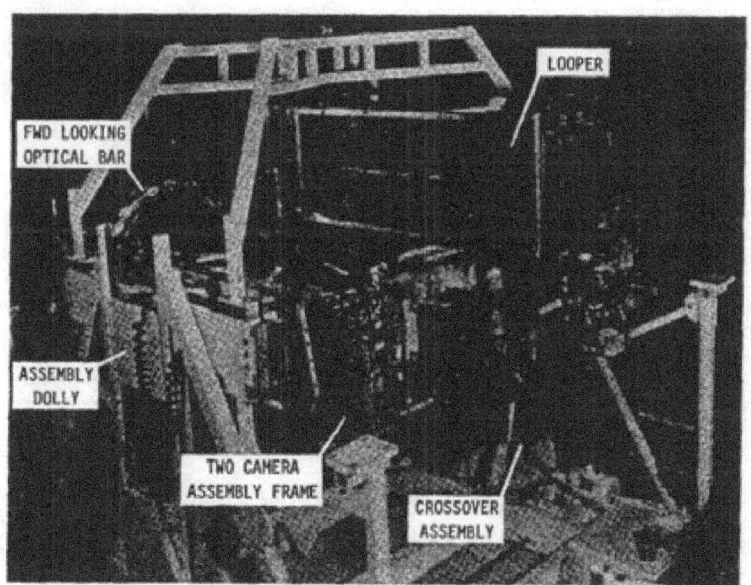

Two-Camera Assembly in Assembly Fixture (SV-17 through SV-20 Configuration)

The Sensor Subsystem Two-Camera Assembly, located in the satellite vehicle midsection, contains a pair of panoramic cameras mounted in a frame. One camera looks forward of the satellite vehicle (Camera A, port side) and the other looks aft (Camera B, starboard side). Each camera has a 60-inch focal length, f/3 folded Wright optical system. The optical system, which contains both reflecting and refracting optical elements, is mounted in the optical bar. The system's 20-inch aperture is formed by an aspheric corrector plate that corrects for spherical aberration inherent in Wright systems.

Light entering the aperture is folded 90° by the folding flat and reflected onto the 24-inch diameter primary mirror. The primary mirror (20-inch clear aperture) focuses light back through the field group mounted in the folding flat's center hole. The field group, with four refracting elements and a filter, corrects for field curvature and residual chromatic aberration. The system's focal plane is just beyond the last field group element.

Either stereo or monoscopic coverage can be selected. The sensor provides complete stereo ground coverage at a nominal convergence angle of 20°. Scanning is accomplished by continuous optical bar rotation, and scan rate is maintained at $20\pi Vx/h$ for continuous ground coverage and a three percent frame-to-frame overlap at nadir. Scan length is controlled by camera shutters; exposure is accomplished by 32 discrete camera slit widths between 0.080 and 0.910 inch.

The cameras can be operated in any of sixteen scan modes (30° to 120° with center angles 0° to ±45°) as selected by the "Tunity" software, with frame format length determined by the scan mode in use. Scan modes are selected as an in-flight option on a per-operation basis. The selected mode remains constant throughout an operation giving Mission Control a maximum target coverage capability with minimum film wastage.

During photography, the optical bars rotate continuously through 360° to provide cross-track scanning, but photography occurs only during a maximum of

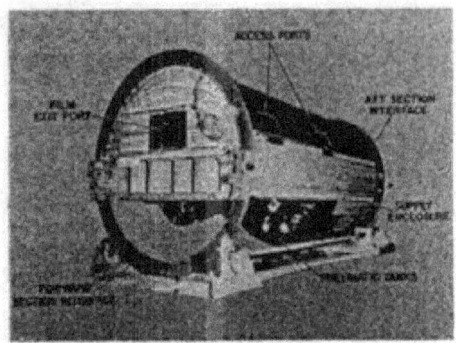

Mid-Section of the Hexagon Camera System

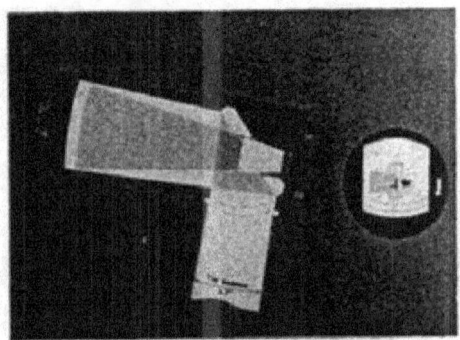

Optical Arrangement of the Hexagon Camera (Folding Wright Optical System)

120° of scan. In each optical bar, a platen (directing the film across the focal plane) is electronically locked to the optical bar through 130° of scan (120° scan plus 10° for settling time, corresponding to the maximum cross-track coverage for the available scan modes) and then recycled to the start of scan position. Platen rotation rate during photography corresponds to optical bar scan rate and is modified for image motion compensation (IMC). A twister assembly guides the film into and out of the platen assembly and accommodates the twisting motion of the film as the platen oscillates back and forth.

Although the in-track and cross-track equations of motion are interdependent, the IMC is independently mechanized in both directions via the platen and metering capstan. The in-track and cross-track IMC signals are generated by the Modulation Computer and are used in the camera's platen and fine film transport system.

Platen and Film Drive Assemblies Mounted on Optical Bar

In the flight direction (in-track), IMC is achieved by the cosinusoidal modulation of the platen with reference to the rotating optical bar during each scan. The modulation is mechanized to compensate for variations of in-track image velocity as a function of Vx/h, instantaneous scan angle ($\emptyset$), the fixed camera pitch angle (θ), and fixed earth curvature. This is referred to as skew angle (ψ) modulation and has a maximum value at nadir of $0.884°$.

In the scan direction (cross-track), IMC is achieved by modulating the predominant film velocity due to scan ($f\emptyset$). The modulation velocity is a function of Vx/h, Vy/h, fixed camera pitch angle (θ), and instantaneous scan angle ($\emptyset$). This modulation is introduced to the metering capstan principally as the integral of velocity, since the servo is positioned-locked to the optical bar during photography.

The following six data records are placed photographically on the film during an operation; latent image start-of-operation marks, latent image start-of-frame marks, scan angle marks, timing marks, satellite vehicle time marks, and forward camera (A) identification marks. The data marks are imaged on the film by flasher write-heads, whose operation are synchronized with appropriate Sensor Subsystem events.

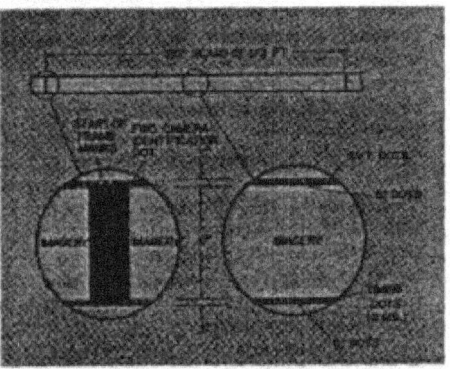

Film Strip Showing Positioning of Edge Data

Operational control of the Sensor Subsystem is provided by servo subsystems controlling the following functions: optical bar rotation, platen drive (photo mode and recycle), slit width variation, coarse film transport (supply and take-up), fine film transport (drive and metering capstan), steerer operation, and platen motion for focus.

Most Sensor Subsystem operational ground commands are processed by a System Command and Control unit that programs the operation of the various servos as required for Sensor Subsystem operation. The Sensor Subsystem is protected from a two-camera catastrophic failure by an independent emergency shutdown design. Critical Sensor Subsystem performance parameters are monitored, and the film transport systems are brought to rest as quickly as possible if the monitored parameters exceed specified limits.

The Sensor Subsystem is organized into subsystems so that most interactions occur within the subsystem; individual subsystems interact as little as possible with one another. The Sensor Subsystem electronic and electromechanical modules are either installed in the electronics compartment, mounted on the two-camera frame, or integrated with subassemblies.

The film path components operating at a nominal constant speed during photography and recycle constitute the coarse film transport system. The forward camera components are arranged differently than the aft camera components because the two optical bar assemblies have a different orientation within the frame assembly. Functionally, however, film moves from supply to take-up in each camera in the same order. Several references in the discussion of the coarse film transport system are included for a complete understanding because of functional overlap between coarse and fine operation and control of film as it travels from supply to takeup.

The distance the film travels from the supply assembly in the aft section, to the first recovery vehicle in the forward section, is approximately 140 feet in both cameras. Throughout its travel over 124 rollers in the "A" camera and 131 rollers in the "B" camera and 6 airbars in each camera, the film must remain centered within specified tolerances.

To correct for displacements of supporting film path elements (i.e., rollers and air bars) caused by structural deformations due to launch and thermal variations, each camera contains active and passive articulators steering the film at critical points in the film path.

Active articulators steer the film across the Sensor Subsystem primary bulkheads (i.e., between the supply and the midsection; and between the midsection and the forward section) to prevent the film from telescoping on the supply and takeup cores. Passive articulators maintain film path alignment between the recovery vehicles and across the two-camera assembly frame in each film path.

The supply assembly maintains film stack integrity in all conditions of powered flight and orbital operations. It supplies film to the two-camera assembly at controlled constant velocities up to 70 inches per second under specified tension and minimizes the potentially large dynamic momentum disturbances inherent in the movement of such a concentrated, relatively elastic mass.

Each take-up assembly, one in each

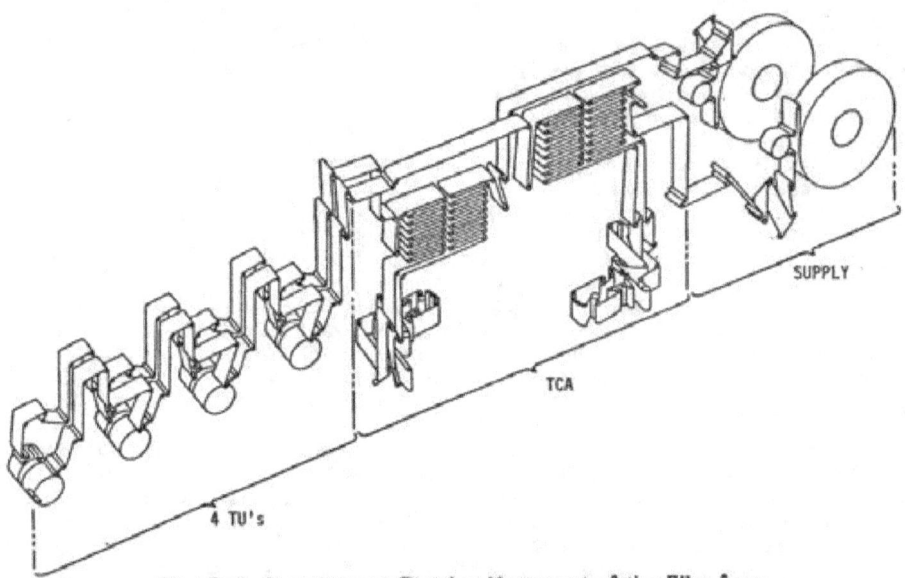

Film Path Arrangement Showing Movement of the Film from the Supply Assembly to the Take-Up Assemblies

Internal View of the Supply Assembly

of the four recovery vehicles, has a film capacity of one-fourth the film load of the supply assembly. Film is pulled from the camera looper by the take-up's drive motor and core.

The looper assembly in each film path serves as the interface between the coarse and fine film transport systems. The looper allows the total length of film stored in it to be constant, but the relative film lengths in the supply and take-up sides of the looper vary as a function of looper carriage position.

View of Looper Assembly

The fine film transport system controls the intermittent film speeds (up to 200 inches per second) required at the camera's focal plane for photography during the various Sensor Subsystem scanning modes. The system consists of the film drive and platen components, input and output drive capstan servos, modulation computer, and system command and control.

The twister assembly, located in the film drive assembly, accommodates the angular change between the rollers in the film drive assembly (which is fixed to the frame) and the rollers in the platen assembly (which is locked to the optical bar during the photographic cycle).

The twister assembly consists of a twin air-bar assembly and a housing that incorporates a manifold through which nitrogen gas is supplied to the air bars. The film wraps one of the air bars prior to wrapping the entrance roller of the platen assembly, and wraps the other air bar after leaving the exit roller of the platen assembly. The twister assembly is free to rotate about its pivot point in response to angular changes between the rollers in the film drive assembly and those in the platen assembly. To accommodate a given angular displacement between these two sets of rollers due to platen assembly rotation, the air bars twist through an angle equal to only one-half of this displacement. This follows from the fact that a given angular displacement (twist) of the air bars results in a corresponding angular displacement of the film path at both the entrance to and exit from the air bars. Using air bars, rather than rollers, the twister permits the film to translate along the length of the bars without damage as the film path twists.

Structurally, the platen assembly consists of the camera's focal plane assembly, slit and shutter assembly, fine tension sensors, P-Mode electronics, IMC position and velocity transducer, and the metering capstan with its motor, encoder and brushless motor electronics. mounted on its own bearings, the platen assembly is located at the focal plane end of the optical bar in the optical bar's inner housing. The platen's outer end, enclosed by the stationary film drive assembly, mechanically interfaces with the film drive assembly through the twister assembly.

Functionally, the platen assembly oscillates on its own bearings independently of the optical bar's continuous rotation, but in synchronism with the optical bar's rotating image of the scanned scene at the focal plane. Its position and velocity, dependent on the optical bar's position and velocity, are controlled by the platen servo external to the platen assembly. As the film, driven by the film drive assembly, moves across the focal plane, the rotating image is exposed on the film with zero smear during the photographic scans. At the end of each optical bar scan, the platen is returned to the start-of-scan position and waits for the optical bar's next scan. This sequence is repeated for each photographic scan.

The capping shutter, located in the platen assembly, opens the camera's aperture at the start of each photographic frame and closes it at the end of each frame in synchronism with the optical bar modes. The shutter's opening blade and closing blade are actuated throughout an operating cycle and repeated for each frame of exposure, consisting of an opening phase, a closing phase, and a reset phase.

The film is completely enclosed in light-tight, pressurized assemblies throughout its passage from the supply assembly to the take-up assembly. The film loaded in the supply assembly prior to launch contains approximately 65 pounds of water defining an effective relative humidity for the film of approximately 40 percent at ambient temperature. The enclosed pressurized film path prevents rapid vaporization of the water from the film emulsion during system operation. Excess vaporization causes two harmful effects: (1) flatness distortion of the film making it difficult to track and producing flutter in the focal plane, and (2) creates a gas layer between film wraps in the TUA causing uncontrolled telescoping as the stack is built up. To prevent excessive water vapor loss, as well as to protect the film from stray light, the film path is enclosed.

The primary (two spherical tanks) and supplementary (one spherical tank) pneumatics systems supply dry nitrogen gas to pressurize the Sensor Subsystem enclosed film path. The systems store approximately 109 pounds of nitrogen under a nominal pressure of 3265 psia at 70°F.

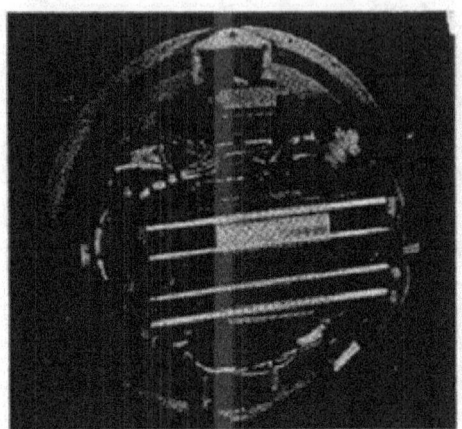

Platen Assembly Showing Focal Plan Rollers

Slit and Shutter Assembly in Test

On-pad, the sealed film path accommodates atmospheric pressure changes through relief and filtered pressurizing valves. When the differential pressure between the film path and SV interior exceeds 0.07 psi in one direction or 0.7 to 0.8 in the other, the appropriate valve will open to reestablish equilibrium below the valve crack pressures. The large relief valves operate in two modes. One uses a small pilot diameter orifice to bleed off small pressure differentials over a long time constant, such as on-orbit or on-pad. The other uses large diameter orifices to effectively "flush" the film path of the excess pressure build-up experienced during powered flight. The valve diameters and response rates are designed to allow no more than the specified differential to exist over the relatively short time span of powered flight.

During initial on-pad and launch venting, the excess gas being dumped is the atmospheric composition experienced on the ground, which will essentially exhibit a relative humidity close to that of the film. As venting continues to flush the initial atmosphere, the water vapor will be reduced. Moisture from the film will, therefore, be given off until the water vapor pressure of the enclosed environment and the film are equal again. Once the correct orbit has been attained, the total film moisture loss due to this effect is less than 1 pound of the 65 pounds carried in the film. In orbit, the allowable leak rate over the mission permits the sealed film path to inhibit further film moisture loss.

As the relief valves open, both primary constituents of the gaseous mixture in the film path or supply (N_2 and H_2O vapor) are bled off. The water lost in this process is replaced by further film outgassing until the water vapor pressure equilibrium is restored. The time constants and absolute values involved in this recurring exchange are small enough so as not to contribute to any detrimental film handling problems.

When film is being transported through the film path, the lower pressure relief valve setting in the film path compared to that in the supply allows a system pressure bleed-off through the vents on the forward steerer enclosure, out a light trap, and hence overboard. Since the supply valve setting is higher, the valve will generally never be required to crack on orbit. During launch, the gas from the supply is dumped to the interior of the supply compartment and ultimately vented to space.

No adverse effects accrue to the system during launch/boost when the supply is venting to its own compartment and subsequently to space. The environment defined in the compartment surrounding the film path enclosures reduces the possible conduction of heat to the film path. If an excess pressure build-up in the film path were vented into the compartment, increased pressure would increase the possibility of thermal conductance between the outer vehicle and the film path. The forward section enclosure is superinsulated and would be equally susceptible to potential thermal conductance if the film path gas mixture were vented forward. The film path vent valves, therefore, dump directly overboard.

Nitrogen for the film path air bars is stored under high pressure and delivered to the air bars at a pressure reduced to approximately 3.35 psig at flow rates of 0.193 to 0.600 scfm (sea level) or 0.193 to 0.300 at orbital altitude (≈ 0.0175 lbm/min). Provision is made within the system for monitoring out-of-specification pressure conditions at select junctures of the flow paths. The "cushion" provided by the air bars is, in effect, a gas bearing over which film passes. This surface must exhibit relatively uniform dimensions at all times that film is moving during camera operation. Nitrogen used for the supply assembly air bars is identical to the film path air bar use. Nitrogen is also used for the supply assembly air bars, seal doors, and brakes.

FIRST FLIGHT OF THE BIG BIRD

It was 15 June 1971, preparations had been completed at Perkin-Elmer in the Danbury facility to monitor the launch and the first flight of the Hexagon reconnaissance camera (unofficially called the "Big Bird") scheduled for launch that day. Arrangements had been made to receive the real-time countdown in the Flight Operations Room (also called the "War Room"). The walls were covered with data boards listing the various characteristics of the Sensor Subsystem (Serial Number 3) on that particular flight.

Charlie Bryant, Manager of Field Operations on the Hexagon program, entered the "War Room" at 8:00 A.M., EDT.[1] Countdown at the Vandenberg Air Force Base on the West Coast was already in progress, having started at midnight (PDT).

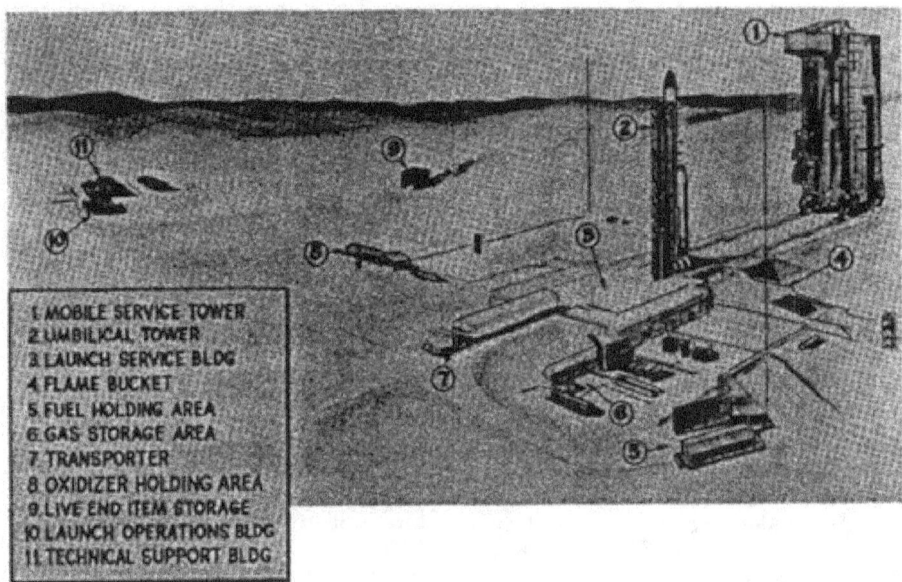

Titan IIID Aerospace Vehicle with Hexagon Payload

The day before, the launch certification form had been signed by W.C. Cottrell, Field Operations Manager on the West Coast, and countersigned by ▓▓▓▓ ▓▓▓▓▓▓▓ customer representative from the West Coast Project Office.[2]

However, Danbury was not the only place where Perkin-Elmer project personnel would gather to listen in on the countdown. Field Operation supervisors and technicians were already stationed at the Vandenberg Air Force Base blockhouse; the Satellite Test Center in Sunnyvale, California); and in ▓▓▓▓▓▓▓▓, the home base of the West Coast Field Operation group.

Slowly, the "War Room" began to fill with the project personnel. One more minute to go, and the small room was filled to capacity. The last ten seconds — and all eyes were on the small speaker box. Liftoff! As the group listened to the progress of the launch, it became more animated. The "Big Bird" was on its way — launch time, 11:41 PDT.

When it was apparent that the launch was successful, the room began to empty. Engineers from the System Engineering group remained behind. It would be their job to monitor the mission on a day-to-day basis and compare the actual sensor subsystem flight data with planned data sheets displayed on the walls of the "War Room."

The "War Room", now doubled in size, is no longer filled with the sounds of countdown during the launch of each "Big Bird." However, it still performs the functions started on that first flight — monitoring the flight data of the Sensor Subsystem.

While the primary objectives of the Hexagon mission was to provide high resolution photography over broad areas, the intent of the first flight was to demonstrate functional operation of the system. The Sensor system achieved this intent.

The sensor system demonstrated a functional orbital lifetime of 31 days. At the end of day 31 when recovery vehicle four (RV-4) was separated, approximately 86 percent of the film had been transported and 57 percent of pneumatics nitrogen gas used.³

Photographic imagery was good with the aft-looking camera results being better than the forward-looking camera. The cameras were not set at the best plane of focus at launch. In-flight corrections were made to each camera to minimize the amount of defocus. The forward-looking camera exhibited residual image smear values slightly greater than predicted in the cross-track direction.

The first attempt to move film was made on the fourth orbit of the Hexagon satellite. The sensor system worked properly, the film was aligned within the film path. Steerers, film tensions, take-up and supply summed errors were well within limits. On revolution 8, the Sensor System health check was performed. All Sensor System executed commands were functionally verified.

On revolution 14, fifty frames were commanded and executed, and a total of 458 feet of film was transported. An additional 58 frames were commanded and executed on revolution 16, a total of 533 feet of film was transported. The Sensor System was now considered operational.

Two days after launch, 20 June 1971, RV-1 was ejected from the satellite on revolution 82. Reentry was nominal; however, main chute cone damage prevented aerial recovery. The capsule was recovered from the water with no damage. Impact location was 8.4 miles south and 3.6 miles west of the predicted impact point. RV-1 returned 40,502 feet of film.

The Hexagon camera had been tested in Chamber A at Danbury and Chamber A2 at Lockheed with collimators that projected targets on the film. These small images (about the size of a dime) were the only indication that the sensor would indeed produce pictures. When the film from RV-1 began to roll out of the Versimats at Kodak and one could see literally miles of imagery, the enormity of the achievement began to sink in. One of the NPIC representatives remarked, "My God, we never dreamed there would be this much, this good! We'll have to revamp our entire operation to handle the stuff."

On revolution 179, RV-2 was ejected with a film load of 53,194 feet. Chute damage was noticed on reentry, but aerial recovery was successful. A major portion of the RV-2 heat shield was detected floating in the water and recovered.

Normal mission operations continued with RV-3 until a camera emergency shutdown (ESD) occurred on revolution 315. Diagnostic and engineering tests cleared the ESD and mission operations were resumed on revolution 326 with the 30 degree camera scans inhibited. Operations continued until revolution 405 when reentry of RV-3 occurred with a film load at 92 percent capacity (54,083 feet). Reentry was nominal, however, a main parachute malfunction occurred and RV-3 plunged into the Pacific Ocean and sank.

An emergency shutdown on Camera B occurred during the trim and seal operation RV-3. Diagnostic and engineering tests were executed and camera operations resumed on revolution 422. An ESD occurred on revolution 445 again on Camera B. Tests did not clear the ESD and monoscopic operations were started on revolution 471. A long engineering test cleared Camera B and stereo operations were resumed on revolution 477.

A command system execution anomaly on revolution 492 resulted in a ESD which cleared itself. The command execution error was determined to be a hardware logic error. Pyro battery degradation was noted on revolution 467 and resulted in a decision for earlier recovery of RV-4 on revolution 502 (Day 31) rather than on Day 45 as had been planned. Monoscopic operations were conducted to balance RV-4 prior to recovery on revolution 502. RV-4 was loaded to 44% of film capacity (25,797) with normal reentry and aerial recovery.

Initial evaluation of the photographic quality of the film recovered in RV's 1, 2 and 4 indicated the capability of the Sensor System to provide the specified photographic performance as identified in the Flight Readiness Report.

The ability to transfer film into each of the RV's was demonstrated, as was the performance of the film take-ups in each RV. A total of 147,799 feet of film was returned in the three RV's that were recovered. Momentary stoppages were troublesome but were in each case cleared.

The first flight of the Hexagon system was truly an outstanding success; the harbinger of many more to come. Not only did it demonstrate that the Hexagon system could operate satisfactorily, the sensor photography met the requirements of close look resolution and broad area coverage in stereo!

The successful SV-1 mission could not have been possible without the dedicated and tireless effort on the part of many individuals. It would be difficult to list all the major contributors without inadvertently omitting someone, particularly since every task was a significant piece of the total program. However, at the launch of SV-1, the following individuals held key positions in the Optical Technology Division. M. F. Maguire, V.P., General Manager Optical Group East (and Acting General Manager of OTD); H. W. Robertson, Deputy General Manager; Dr. R. M. Scott, V.P., Technical Director; R. C. Babish, E. B. Brown, C. S. Lapinski, Dr. R. E. Hufnagel, and B. Malin, Members Technical Advisory Board; V. Abraham, Director Advanced Planning; P. E. Petty, Director Program Management; R. W. Jones, Director Engineering; C. Karatzas, Director Assembly and Test; J. Braddon, Director Product Assurance; M. A. Mazaika, Manager Advanced Programs; W. H. Benson, H. E. Henderson, G. O. Henderson, A. Wallace, V. C. Buonaiuto, and R. A. Kelley, Managers Program Management; R. W. Williamson, W. Newell, K. W. Hering, S. T. Karachuk, R. D. McLaughlin, N. A. DeFilippis, L. J. Farkas, P. J. Convertito, J. S. Patterson, J. J. Garrish, R. H. Carricato, R. Labinger, F. Scott, L. B. Molaskey, M. H. Krim, and W. E. Keeney, Managers Engineering and Sr. Technical Staff; K. H. Meserve, C. O. Bryant, W. Cottrell, F. E. Johnson, and T. A. McClung, Managers Assembly and Test.

SENSOR SUBSYSTEM IMPROVEMENTS

Throughout the design of the Sensor Subsystem, changes were made whenever it became possible to enhance performance and/or reliability of the Sensor Subsystem. After the first flight of the Hexagon camera, both the customer and Perkin-Elmer started considering additional improvements. Soon after the flight film from the first mission was evaluated, we began thinking about ways to reduce smear, carry more film, get more coverage and at the same time increase system reliability.

Although the supply was rewound on the first mission, it was soon discovered that rewind had to be restricted because of film-induced tracking problems and tracking problems caused by various kinds of debris entering the film path from various sources in the space vehicle. Film wedge and other mechanical film properties caused mistracking of the film during film transport. Hydrodynamic liftoff due to gas ingestion during rewind led to film spillage in the Supply Assembly. Several modifications were made to accommodate this situation. Fence barriers were placed in the Supply Assembly to prevent film spillage from jeopardizing the other film path and steering was limited so that rewind would be possible at higher speeds.

The problems of film wedge were discussed with the supplier, Eastman Kodak Company. The length of the film strips making up a roll of film were limited and the strips were arranged to prevent the film wedge from accumulating on one side of the film spools thus creating a film taper and causing film spills.

The Sensor Subsystem was originally designed for SO-380 film. Prior to the first flight, all testing was accomplished with SO-380 film. At the customer's request, Perkin-Elmer tested a very nominal amount of 1414 film, and, as a result, film for the first flight was changed from SO-380 to 1414 film. This change increased coverage since 1414 film was thinner than SO-380. However, the film taper first observed with 1414 film contributed to tracking problems.

By mission SV-4, the customer asked Perkin-Elmer to add color film (SO-255) to the film supply to enhance the capability of the Sensor Subsystem. This required an in-flight filter change which necessitated the addition of a filter mechanism in the Optical Bar Field Lens Assembly.

During the first six flights (Block I), the customer and Perkin-Elmer program management began to consider improvements for the Block II Sensor Subsystems. Since it was determined that the film stacks in the Supply Assembly were subjected to lower launch vibrations than anticipated, it was decided to eliminate supply caging. After several months of analysis and ground testing, the supply caging device was removed from the flight models. This resulted in fewer parts and less weight, thereby increasing reliability. With the elimination of the supply caging and less demand on the nitrogen gas, the way was clear to expand the size of the film supply spool. Perkin-Elmer also added a $180°$ builder roller in the Takeup Assemblies incorporating the change on SV-9. The new $180°$ builder roller in the Takeup Assembly improved tracking stability and thus accommodated the residual film wedge and film crowning in the film rolls. During the addition of the builder roller in the Takeup Assembly, the takeup was modified to carry a larger film roll, thus paving the way for increasing the film supply diameter. As a result of various system improvements, the resolution of the sensor subsystem improved significantly (see Appendix E).

Three major improvements were made in Block III: the capacity of the nitrogen supply was doubled from 34 pounds to 68 pounds; the Solid State Sensor (S^3) Camera replaced the Itek Stellar Cameras and mensuration changes were made; and the Large Looper was added to decrease interop wastage thereby increasing the quantity of imaged film by about 20%. Film resolution was also improved by replacing the 1414 film with SO-208 film. In addition to a change to thinner film, the 66.6-inch diameter supply film roll was increased to 68 inches in diameter.

Throughout the production run of the 20 Sensor Subsystems, a large number of design changes were made to increase reliability to accommodate the continually increasing length of the missions (from the original 30 days during SV-1 to 261 days during SV-16).

One of the most significant changes made to the Hexagon Camera was the addition of the S^3 cameras. Beginning with SV-5, the Defense Mapping Agency flew the APSA Camera (Itek Stellar camera) which photographed the star field and the ground simultaneously. This arrangement permitted the DMA to correlate the attitude of the vehicle with the terrain and, from that, arrive at the target location. As time went on, the DMA requirements became more stringent. The DMA begain to use the Hexagon imagery for their mapping requirements. The imagery was transferred from the panoramic film onto the APSA camera data and then final measurements were made on the APSA camera film. Since this was a tedious process, the DMA asked the Hexagon Program customer and Perkin-Elmer to consider converting the Hexagon Camera to a Metric Camera.

The government funded many studies with various companies to determine if panoramic photography could be used for mapping. As a first step, the government developed a device called the GEOPAC which was used in conjunction with the Hexagon camera on the SV-15 mission. In addition, Perkin-Elmer calibrated the optical bar encoder, changed the $5°$ scan marks to $1°$ scan angles and changed the configuration of the fiducial marks from dots to fine cross hairs so that a more accurate determination of film shrinkage could be made. The results prompted the DMA to ask for serious consideration of making the Hexagon camera a metric camera.

The major problem, however, was locating a stellar camera on the two-camera frame to establish an accurate interlock angle between the stellar camera and the optical bars. Several alternative designs were considered by Perkin-Elmer; however, they all proved to be too complicated and risky to the primary mission of the Hexagon Camera.

Lockheed, an Associate Contractor on the Hexagon Program, came up with an idea of placing a gyro package and tying it into the Sensor Subsystem. However, the accuracy achieved in this way would be about ten arc-seconds and was twice the system requirement. Victor Abraham, presently Hexagon Program Director, believed that the idea of using the Hexagon Camera as a Metric Camera was in danger of being abandoned. "It was then that I realized," said Abraham during an interview, "that the Charge Coupled Device was the only solution to the problem."

Victor Abraham, Hexagon Program Director

Abraham had previously worked at the Fairchild Camera Company as a physicist developing CCD devices for a variety of applications. He called a meeting with Dr. Roderic Scott, Technical Director at Perkin-Elmer, and Michael Weeks, OTD Division Manager, and suggested the idea of using CCD's in a Star Sensor System. Dr. Scott immediately saw that it was a perfect solution to the problem since it could be designed into a small package using small optics and would not require a film transport since the star data could be processed in digital form.

Some conceptual work was produced by program engineers showing that the S^3 camera was feasible. A series of presentations were made to the government, and permission was granted to pursue the idea. Perkin-Elmer had to prove that not only would the S^3 camera achieve the necessary accuracy of five arc-seconds but would in no way endanger the primary mission of the Hexagon Camera.

The S^3 project was initiated and a CCD characterization laboratory was constructed. Analyses were conducted and an S^3 camera was built and tested. It was flown in SV-17 and met all mission requirements. Although the S^3 camera suffered a major anomaly which made half the S^3 system inoperative, the camera exceeded the 5-arc-second requirement (3½ arc-seconds). The Hexagon Camera is now called the Metric Panoramic Camera System. Vic Abraham, who promoted the S^3 camera, noted in an interview that, "The Hexagon Camera has become a very versatile instrument and can now be used not only for intelligence-gathering operations but also for mapping."

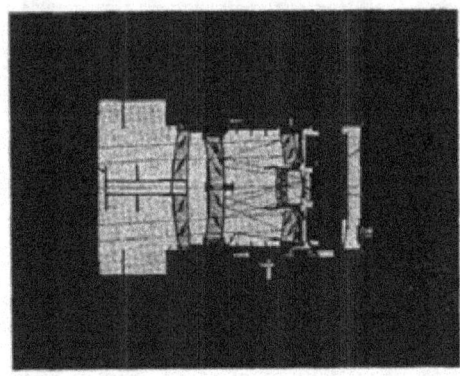

Solid State Sensor (S^3) Camera

In addition to the S^3 camera, the SV-17 mission carried one other major improvement - a large looper with an increased storage capacity. The capacity of the original looper was 10 feet of film.

MAJOR HEXAGON SYSTEM IMPROVEMENTS

Improvement	Effect	Operationally Effective
OOAA Command Assembly	On-orbit image motion compensation	SV-4
Smear Slits	Engineering tool for measuring on-orbit IMC errors allowing corrections to be implemented	SV-6
In-Flight Changeable Filter	Change filter during flight to accommodate color film	SV-7
High Capacity Pneumatic Supply	Provided greater capacity for film transport and longer operational life	SV-11
Large Diameter Supply	Increased film supply capacity	SV-12
UUTB Film	Increased film supply footage	SV-14
Mono Cubic Dispersed Emulsions	Improved photographic system performance (best measured GRD to date ▓)	SV-15
Large Looper/MFT	Decreased film wastage by approximately 20%	SV-17
S^3 Camera System	Provided pan system with metric capability	SV-17
Supplemental Pneumatics Supply Module	Provided additional capacity for film transport and longer operational life (longest mission to date - 270 days)	SV-17

However, because of the problems which occurred on the first few flights during the rewind operations, it was decided to restrict the rewind operation. This, of course, resulted in film wastage. In this mode, the looper served only as an interface between the coarse film path and the fine film path and was not used for storing film during the end of the run. It was apparent that something had to be done to correct this situation.

However, it was not until Block III that the customer and Perkin-Elmer began to seriously examine the possibilities of developing a larger looper and thus eliminating the necessity of a film rewind onto the supply. Leonard Farkas, who at that time was in charge of the System Integration Department, started an analysis to determine the optimum film storage capacity of a larger looper. It was determined that a 40-foot capacity was the optimum size. This would enable the looper to store sufficient film for a 90° scan, with a margin of safety.

A large looper was built and tested on the Development Model and flown successfully in SV-17. Prior to SV-17, film wastage (unimaged film) was 24% because of interframe space and interoperation space. The large looper recovered 90% of the wasted film.

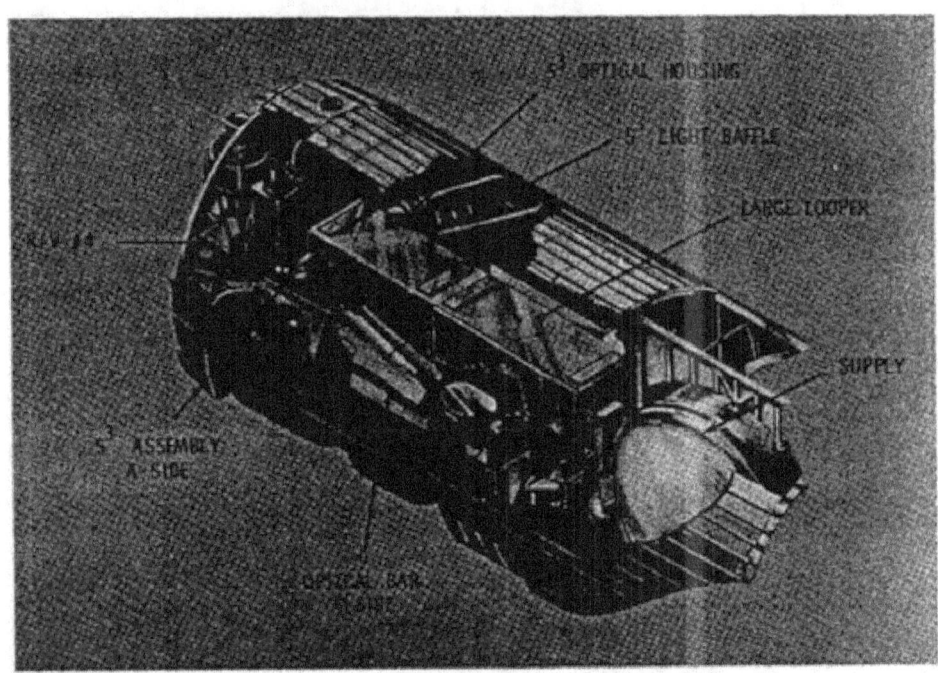

Major Elements of the Metric Panoramic Camera System

2 CUSTOMER RELATIONSHIPS AND INTERFACES

PROGRAM MANAGEMENT

In 1964, when Perkin-Elmer first became involved in space reconnaissance programs, both the Central Intelligence Agency and the United States Air Force were conducting separate space reconnaissance programs. In time, the roles of the Agency and the USAF were defined by the government, but it was the climate in which the Hexagon program was born.

Perkin-Elmer was approached almost simultaneously by both the Agency and the USAF to work on study programs for a space reconnaissance system. In the beginning, the Agency and USAF requirements were different enough so that there was no conflict of activities at Perkin-Elmer. But in a few short weeks, it became apparent to Perkin-Elmer management that the goals of both programs were converging. Since it was the Agency that first approached Perkin-Elmer to work on the reconnaissance program, Perkin-Elmer management informed the USAF that it could no longer work on their program since the government would then be paying twice for the same effort. After a meeting with Dr. Brockway McMillan, Under Secretary of the Air Force, Perkin-Elmer was relieved from participating on the USAF program.

At the time that the Agency approached Perkin-Elmer, John A. McCone was Director of Central Intelligence. McCone recognized the importance of space reconnaissance systems and, soon after he became Director in 1961, he attempted to assert the CIA's leadership position in this area. He created the Directorate of Science and Technology and recruited a brilliant young scientist, Dr. Albert D. "Bud" Wheelon to head the directorate. The directorate was composed of several offices. The Office of Special Activities was responsible for the development and control of all reconnaissance programs in the CIA.

In 1965, the Office of Special Activities was split into two offices; the Office of Special Activities and the Office of Special Projects. The Office of Special Activities continued to control all aircraft reconnaissance programs, but all space reconnaissance programs were transferred to the Office of Special Projects (now called the Office of Design and Engineering). John S. Crowley became the first Director of the Office of Special Projects.[1]

It was in this office that the requirements for the Hexagon Sensor Subsystem were formulated. The forerunner of the Hexagon program, the Fulcrum program, started in January 1964 when the Agency selected the Itek Company to begin initial studies of a camera system embodying the capabilities of both the area-surveillance and the close-look satellites already in operation.

In June 1964, the Agency contracted with Perkin-Elmer for studies paralleling the activities at Itek. This technique is often used on critical government programs to insure the development of the most effective system.

In September 1964, the Perkin-Elmer study was expanded into a Phase I study and resulted in the construction of a camera mockup called the "cocktail shaker." It was soon after that the Itek Company withdrew from the Fulcrum Program.

The Agency asked Chester W. Nimitz, Jr., if Perkin-Elmer could continue the activity started at Itek. Nimitz accepted under the condition that Perkin-Elmer would have an opportunity to study the Itek concept before undertaking a final design. The results of these studies were reported previously.

Relationships between the Agency and Perkin-Elmer program management at this point in time were excellent. However, as in any endeavor or partnership involved in a critical undertaking involving the national security, the pressures of designing a unique camera system within a tight budget and schedule began to strain this good relationship and soon, after the award of contract, "hairline" cracks in this relationship began to appear in several areas simultaneously. One area concerned the requirement of customer approval of all design decisions; another involved customer approval of all subcontractor selections; and the one that seemed to have the greatest impact on the cost and schedule of the Hexagon program was the involvement of the Agency's technical consultant, known as SETS (System Engineering Technical Staff), in all technical and planning decisions.

At the time that the Agency contracted with Itek on the Fulcrum program, it also contracted with the Thompson-Ramo-Woolridge Corporation (TRW) to provide technical consultants to review Itek's progress on the program. In addition, TRW also worked as the System Engineering Assembly Checkout (SEAC) associate contractor on the Fulcrum Program. Itek management objected to the dual function that SEAC served on the program. SEAC's dual function continued when the Agency transferred Itek's Fulcrum activities to Perkin-Elmer in March 1965. When the Hexagon program contract was awarded to Perkin-Elmer, the TRW technical consultant group then became known as SETS (System Engineering and Technical Services Contractor). Concern for the working relationship between the Agency, SETS, and Perkin-Elmer was expressed in a memorandum from John Crowley to Carl W. Besserer who headed SETS.[2]

When John Crowley met with Chester W. Nimitz at the end of 1966, he expressed a serious concern with Perkin-Elmer's ability to properly staff the Hexagon program.[3] Nimitz decided to transfer the management of the program to Kennett W. Patrick, then General Manager of the Electro-Optical Division (EOD). Patrick replaced Dick Werner as General Manager of OTD on 1 January 1967, bringing with him a substantial number of EOD technical and administrative personnel.

This move satisfied John Crowley until the fall of 1967 when in a letter to Nimitz he expressed concern with the program's progress and indicated that decisive action was necessary to recover both technically and schedule-wise.[3]

Nimitz responded by a division reorganization and a redirection of the program assets, however, Crowley's letter disturbed Nimitz prompting him to reply and defend Perkin-Elmer's record on the Hexagon program.[4]

As the program progressed, the increased number of customer representatives and SETS personnel assigned to the Perkin-Elmer facility began to cause difficulty at the working levels of Perkin-Elmer. Perkin-Elmer engineers and administrative personnel began to accept verbal suggestions of both customer and SETS representatives as official direction to make changes. Unfortunately, these changes affected not only the cost of the program but also the schedule.

Toward the end of 1968, the Agency was facing difficulty in acquiring sufficient funding to support the Hexagon program at the increased level of activity required to maintain schedule. Perkin-Elmer responded to this situation presenting a reformatted program which permitted a substantial cost reduction.[5]

Delivery of the first flight sensor was 48 months after receipt of the development contract for six vehicles (Block I). This remarkable achievement was accomplished within four months of contract and with less than 25% overrun. The Agency reacted by retroactively changing the contract for SV-2 and subsequent to an incentive contract including cost, schedule, and on-orbit performance. This unique arrangement served both the Intelligence Community and the Contractor exceedingly well. By placing incentives on those factors that were important to the Customer, the Contractor was motivated to strive for optimum results. This was one of many pioneering approaches implemented on the Hexagon Program that contributed to its outstanding record of success.

CUSTOMER CHANGEOVER

In December 1971, six months after the first successful Hexagon mission, the government notified Perkin-Elmer of its intent to transfer responsibility for the Hexagon program from the CIA to the

USAF, specifically the Air Force/Secretary of the Air Force, Special Projects Office (AF/SAFSP). It was subsequently determined that the effective date of this action would be 1 July 1973. However, there would be approximately a one-year overlap by these organizations. The CIA would continue the Block I and II programs until that date, but the AF/SAFSP would begin the procurement of Block III hardware immediately.

In May 1972, SAFSP issued a request for quote covering production of Sensor Subsystems 13-18 (Block III) and associated services, and a contract was negotiated in October.

The USAF's stated philosophy of requiring the contractor to manage the Hexagon program without the use of a System Engineering Contractor for technical and administrative decisions was in contrast to the management techniques used on the initial Block I and II contracts. It should be noted, however, that the SAFSP Office maintains very close surveillance of the program and participates in all important decisions.

The philosophical difference of the Block III contract broadened Perkin-Elmer's responsibility and control of the program. A more general statement of work was written and more conventional contracting was negotiated. The "Martin" formula for contracting for special satellite programs was followed. This unusual procedure was evolved during General John Martin's tenure as SAFSP. The Block III contract followed both the spirit and intent of the Martin formula. One only has to look at the program results to judge the efficacy of this contracting method.

The transition of a program as complex as the Hexagon sensor from one government agency to another in midstream was indeed an unusual and innovative procedure. That it happened without missing a beat is a tribute to the professionalism of the individuals involved on all sides.

PROGRAM SECURITY

A major concern of the CIA and the SAFSP is program security. A review of the Hexagon program TWX messages from 1964 reveals that although all aspects of security are carefully monitored, there are several areas of specific concern; indiscriminate use of insecure telephone systems,[1] release of information by the press concerning visible changes to the contractor's organization and facilities related to the program, and program personnel reaction to newspaper stories, magazine articles, and books revealing various aspects of satellite reconnaissance.[2]

Shortly after Brigadier General John E. Kulpa, Jr. became Director of the Office of Special Projects, Department of the Air Force (August 1975), he wrote a letter to Paul Petty (General Manager of OTD) reaffirming the basic policies which governed their business relationship.[3] His letter highlighted areas he felt deserved special comment and emphasis including contract management, marketing, and ethics. He also reiterated the importance of security and stated the following.

"Security requires constant attention. Satellite intelligence collection systems are vital to the nation -- and extremely vulnerable. This vulnerability not only includes the threat of physical damage, but extends into potential political countermeasures. The Byeman control system serves to prevent unfriendly nations from finding reasons to exert political pressures against our reconnaissance satellites, as well as protecting their true capabilities and the military and industrial base engaged in their development and operation. Adversary nations are known to be taking dedicated, effective countermeasures, including warning alerts, expanding use of information encryption, cover, camouflage and deception, to reduce the effectiveness of our reconnaissance satellites. We cannot afford to give them any advantage.

The recent public revelations of various aspects of satellite reconnaissance have been unfortunate. However, we must continue not to comment on printed stories nor openly discuss this subject. We each need to reaffirm to all of our people that the policies and procedures of the BYEMAN control system must be followed in spite of greater public awareness; if, in the future, there are changes in security policy, they will be conveyed to you directly and quickly from me."

One of the magazines continually publishing articles on the progress of this nation's reconnaissance programs is AVIATION WEEK.[4-24] Other publications have sporadically covered this area of government activity, but the most prolific writer on the subject has been AVIATION WEEK's Philip J. Klass.[25-43] He did a thorough job in summarizing his knowledge of the United States reconnaissance program in a book titled "Secret Sentries in Space.[44] Although lacking in some details, and slightly inaccurate in others, little is left to the imagination of our nation's adversaries. What is really damaging is the impact these public revelations have on people involved in reconnaissance programs. After each public announcement of the newest advances in this country's space reconnaissance activities it is necessary for program management in both government and industry to caution their personnel not to discuss or confirm these stories.

How this information is obtained is a mystery since security officers both at Perkin-Elmer and the government deny that they are controlled leaks. If this is true, then it must be assumed that all of the published information on our country's reconnaissance programs is gathered from people in government who have knowledge of the programs and are trying to gain political advantage; or revealed to reporters by embittered employees who have been terminated from the program; or obtained from ex-government intelligence agents running a crusade against our nation's intelligence community and making a "few pieces of silver" in the bargain; or foolish people working on the program who gain some measure of importance by telling a reporter "something special" known only by a few.

Certainly, the bad press that our security forces have received in the past 20 years has had a deteriorating affect on the general security attitude of the citizens in this country, thereby weakening our resolve and ability to gain intelligence and develop an effective response to our adversaries' thrusts. Fortunately, the pendulum is now swinging in the other direction and the legislation of the 60's and 70's, which made it extremely difficult for our security forces to operate, is now being modified.

Prior to the customer changeover on 1 July 1973, SAFSP conducted an audit of Perkin-Elmer's operations as related to the Hexagon program.[45] One of the areas covered was program security. Fact-finding sessions were conducted during the week of 11-15 June 1973. The sub-team responsible for reviewing the security organization and policies conducted an extensive examination covering all aspects of program security.

In a letter to Robert H. Sorensen, President of Perkin-Elmer, Brigadier General David D. Bradburn, who at that time was Director of the Office of Special Projects, stated, "I am very pleased with Col. Parrish's initial report to me on the conduct of the response to the survey team and the demonstrated professionalism of your top managers in the OTD. I believe there was candor and frankness all around and that my managers and yours developed excellent rapport. After reviewing the survey report in detail I will advise you of the significant conclusions and any actions I think you should take."[46]

In a subsequent letter containing a summary report on the management survey at Perkin-Elmer, Bradburn included a paragraph on security and suggested some actions.[47] "I place much more dependence on your security staff than the CIA did. I look to your security man to handle security planning and staff functions and to handle most security problems through direct and frequent contact with my security staff here in Los Angeles. This arrangement makes your security officer an extension of my office with considerable discretionary authority. He needs to be involved in every aspect of program management and hence should report to the Hexagon Program Manager."

Some basic differences between CIA and SAFSP security direction were also noted in the Bradburn letter. "There is no integrated Byeman security structure within Perkin-Elmer. There is no principal

Byeman security officer within the Perkin-Elmer corporate structure. The Corporate Security Office is located within the Electro-Optical Group. There is no organizational relationship between security functions. At present, Optical Technology Division Security is (physically) located in the Product Assurance Department -- a situation which is anomalous to the management philosophy of SAFSP. There is little incentive for top quality Byeman security management personnel to remain with the organization. The internal Byeman security systems and procedures in effect at the OTD to support the Hexagon program are generally outstanding. This is due largely to continuous on-site supervision and direction by a resident security representative of the CIA security staff, and to the competence of the OTD Security Group.

The significant change in security operations is that there will no longer be a customer resident security representative, and SAFSP looks to the contractor to motivate his people and resolve more problems on his own. To accomplish this, a firm rapport must be effected. To insure that the present program does not lose its effectiveness, the principal Byeman security representative of the division must be positioned sufficiently close to the top of the management structure where he can function as an integral member of the management team and effect a competent, cohesive interface with SAFSP."

The letter indicated that the cover and Phase II briefing material and presentation were good, although Perkin-Elmer was to stress more definitely that the "fact of" satellite reconnaissance is a Byeman secret and any direct association between optical/camera operations and military space must be handled at the Byeman level. The difference between Byeman security and normal DOD classified information was also to be emphasized. Under SAFSP direction, Perkin-Elmer's Byeman security staff was to assume the responsibility of giving Phase III level briefings, a function previously handled by the resident CIA security officer. The Bradburn letter ended with the statement that, "All other plans and all actions in this area (security) are good."

In a letter to General Bradburn, Robert Sorensen listed the actions that were taken in response to the SAFSP Management Survey Report. The Optical Technology Division was reorganized and Paul Petty was appointed Deputy General Manager of OTD with full-time assignment as Program Manager of the Hexagon Program reporting to Michael Weeks who in April 1973 had replaced Michael Maquire as General Manager of OTD. B. Alan Ross became Assistant General Manager, OTD, and assumed responsibility for Security, Contracts, and the functions previously designated as Program services.[48] These moves satisfied SAFSP management.

Although the survey report stated that "All elements of physical security are outstanding," the report also noted that, "Contrary to the contractor's (Perkin-Elmer) presentation, not all personnel working in OTD require Hexagon access. Specifically, personnel working on company proprietary or commercial activity in Building 3 of the Danbury building complex do not have automatic "need-to-know." This was the first indication that the mixing of Byeman work and other activities created a security problem. The Danbury OTD complex, which appears to be one large facility (370,000 square feet), is actually five separate buildings within one exterior shell. A new addition, housing both EOD and OTD people, was added on to the Danbury complex in 1980.

On October 1977, Perkin-Elmer was awarded the contract for NASA's Space Telescope. It had previously obtained permission from SAFSP to use part of the Danbury OTD facility for the Space Telescope. The building was originally built for and dedicated to the Hexagon program.[49,50]

Prior to the inclusion of the Space Telescope program, the OTD Danbury facility was primarily a Byeman facility with few non-Byeman people. This changed when Space Telescope personnel arrived on the scene. The cafeteria, the hallways, and the building grounds were now being shared with non-Byeman people and Hexagon Program personnel now had to exercise additional caution to safeguard program security.

This arrangement was working and the security of the Hexagon program was intact. However, on 22 April 1981 an incident occurred that was to precipitate a strengthening of the Hexagon security program.

On Thursday morning, 23 April 1981, 0720 hours, a Perkin-Elmer employee noticed two Hexagon shipping containers located outside of the Danbury building near the entrance to the gross cleaning and shipping area. He immediately rounded up Byeman-accessed manufacturing personnel and moved the Hexagon shipping containers inside the gross cleaning area. It was determined that no penetration was made on the boxes of swath material and shipping containers, and the Security Office was notified of the violation.

Hexagon Film Shipping Container

As a consequence of this incident, SAFSP conducted a security review at the Danbury OTD facility. The conclusion reached during this review by the SP-3 audit team was that the Perkin-Elmer Byeman Security Program was not entirely compatible with the current requirements of the Byeman Control System.

A major effort was launched to correct the discrepancies discovered by the SP-3 audit team. The OTD Security Department was reorganized and placed under the management of Sheldon Ferber and a massive Security Awareness Program was established to rebrief all Byeman accessed personnel at Perkin-Elmer. The perimeter of the Byeman area was modified to make it more secure and the communications and computer systems were separated from the "white" areas in the Danbury building complex.

These actions resulted in a Byeman facility that is greatly strengthened to prevent penetration from the outside.

Mixing Byeman and non-Byeman programs in the same building complex creates problems for security personnel. However, the cost savings to the government are substantial. Government-owned equipment can be used on non-Byeman programs and critical skills needed on Byeman programs can be retained by assignment to non-Byeman activities.

Byeman secrets can be protected by an alert and creative security staff supported by all levels of management. Byeman program personnel can be motivated and trained to adhere to the stringent requirements of working on a Byeman program surrounded by non-Byeman activities.

A continuing program of security awareness is critical to Byeman programs since the media delights in airing our Nation's secrets. An excellent example of what Byeman program personnel are subjected to was viewed on network television on 20 July 1975. The program "60 Minutes" reviewed United States progress in space reconnaissance. In a 17-minute segment, Mike Wallace interviewed Philip J. Klass (a senior editor of AVIATION WEEK), Michael Marchetti (ex-CIA agent), and General Lucius Clay, Jr. (Commander of NORAD).

In his closing statement, Mike Wallace said, "If you think we've just revealed space secrets to the Russians be advised they know all about building 213 (National Photographic Interpretation Center) and a lot more. Then why all the secrecy? Well, some people in our State

Department say its so as not to embarrass the Russians. They don't want to have to admit to their own people that U.S. satellites are photographing the Russian land mass day in and day out. But that doesn't seem to make much sense because, after all, the Russians are doing the same thing to us. Others suggest the reason for the secrecy is so as not to inflame other nations because these satellites can, of course, photograph any spot on earth. And, of course, these other nations know all about the satellites but they've never been publicly confronted with them. If they were, the fear is they would have to make a fuss about it in the United Nations, and neither the Americans nor the Russians are anxious for all that. If all this sounds like something out of "Alice In Wonderland," you're right."

THIS PAGE INTENTIONALLY LEFT BLANK

3 TECHNICAL DESIGN, MANUFACTURE AND TEST

EVOLUTION OF THE SENSOR SUBSYSTEM DESIGN

The evolution of the Sensor Subsystem design recommended in the Perkin-Elmer Hexagon proposal began in March 1965 when Perkin-Elmer was funded by the Agency to continue a camera design initiated by Itek. Perkin-Elmer engineers studied the Itek "optical bar" design to determine if that particular configuration was the best technical approach. In August 1965, Perkin-Elmer recommended to the Agency a modified form of the "optical bar" design. From that point in time to the day Perkin-Elmer was awarded the Hexagon contract for the Sensor Subsystem in October 1966, Perkin-Elmer engineers conducted studies and experiments to develop a film transport and electronic controls for the "optical bar" design. Dr. Roderic Scott attributes the selection of Perkin-Elmer by the Agency to: (1) our willingness to work with the CIA, (2) the design of the twister, and (3) the demonstration of a process to manufacture high quality optical flats at a rate of one per month.

When Perkin-Elmer first became involved in Phase I of the Agency's reconnaissance program in September 1965, system engineers initiated a Preliminary Performance Specification based on oral and documented requirements from the Agency. This specification was updated as newer requirements evolved during the various study efforts on the program. It was incorporated into Perkin-Elmer's Hexagon proposal submitted to the government in July 1966 and eventually became the Sensor Subsystem Specification (SP-621-0001) referenced in the Hexagon negotiated contract.

When the Hexagon contract was awarded to Perkin-Elmer in October 1966, the design and development of the proposed system, supported by Agency funding, was already in progress. In many areas conceptual layouts were completed and detail design layouts were well underway. Little time was lost in starting up the program since Perkin-Elmer had spent 18 months prior to the award of contract developing and testing the proposed design concept.

The System Engineering Section of the newly organized Optical Technology Division subdivided the Sensor Subsystem into functional units and, based on the Performance Specification design criteria, established error budget allocations for each design area. The total of the error allocations included all factors which influenced the quality of the image on the film.

Project engineers in the various engineering sections (i.e., Optical, Mechanical, and Electrical) responsible for designing these units were directed to prepare a functional specification for each design area. The functional specifications were to be the primary guidelines and contained system engineering data, initial performance and environmental requirements, design criteria, and all interface requirements or design constraints known at the time the specification was prepared. The development of the functional specification was an iterative process and required close liaison between system engineers, project engineers, and program personnel responsible for interface control.

As the conceptual design of each functional unit was completed, the project engineer presented it at a technical review attended by customer representatives and consultants and Perkin-Elmer's Technical Advisory Board. The purpose of the Concept Review is to judge the adequacy of the design and to uncover any technical weaknesses or interface problems. It was the initial milestone in the program schedule each functional unit had to pass before continuing to the next design level. The Concept Review Data Package supporting the design approach included an approved functional specification, design layouts, and engineering study, analysis, and test reports conducted on the design.

Following approval of the Concept Review, the project engineer and his design team prepared for the next level of technical review, the Preliminary Design Review (PDR). The first step was to prepare a design specification for the functional unit documenting engineering requirements for the design and performance of the unit and outlining the methods for verifying the design. After completion of final design layouts, the design team prepared a PDR Data Package to support the design approach. In addition to a complete set of design layouts, the data package included an approved design specification, and engineering analyses and test reports supporting the final design. The project engineer updated his design to conform to any suggestions or changes directed by the customer or the Technical Advisory Board, after which customer approval was granted.

There was one last technical review each functional unit had to pass successfully before the design could be transformed into manufactured parts, the Critical Design Review (CDR). The CDR Data Review Package consisted of a complete set of manufacturing drawings including detail, subassembly, final assembly, specification, and source control drawings and parts lists. In addition it included an equipment specification for the functional unit documenting the configuration and performance during the fabrication or production phase and specifying the inspections and tests required to verify the equipment. This was an extremely important review since any changes made after the drawings were released to the manufacturing and purchasing departments would have a serious impact on both costs and schedule.

In addition to the functional units, this sequence of technical reviews was applied to all test equipment, assembly and test stations, and aerospace ground equipment required for the support of the Sensor Subsystem.

This section of the history will review the evolution of the functional units and the impact that the basic design problems and their solutions had on the Hexagon program. Since only a brief technical description of each functional unit will be provided, the reader is asked to refer to the Technical Data Book for a detailed design description. Important changes affecting particular units are discussed as each functional unit is covered. In retrospect, the engineering decisions made throughout the evolution of the Sensor Subsystem design were timely and correct and contributed to the successful completion of each mission.

Three significant problems had to be solved to successfully produce a Sensor Subsystem capable of meeting the requirements of the Hexagon Reconnaissance System. In addition to handling large quantities of wide (6.6 inches) ultra thin-base film at a high velocity and to a required accuracy of synchronization and achieving an acceptable environment for the film, it was necessary to design lens and mirror mounts capable of maintaining optical alignment through launch and in orbit. Also, new grinding and polishing techniques had to be developed to produce the significant quantities of large, high-precision optical elements required to support the Hexagon program.

Optical Bar Assembly

The Optical Bar Assembly is a completely enclosed and rigid structure containing the camera optics. Its three primary purposes are to provide a mount for the optical elements and isolate them from external mechanical deformations; provide thermal protection for the optical system; and provide the basic rotating motion for the transverse scan and the system's primary time reference.

The "optical bar" design was started by Itek in the spring of 1964 and continued by Perkin-Elmer in March 1965 at the request of the Agency. After a critical review, the "optical bar" design was improved and recommended in Perkin-Elmer's Hexagon proposal.

At the award of contract, the design of the Optical Bar Assembly was further along than other functional units since the long-lead time required for the optical elements required the release of drawings early in the program. During the first eight weeks of the Hexagon program several optical design changes were evaluated. One modification increased the half-inch back focal length to one inch to provide sufficient space for a focus sensor, a capping shutter, and a timing dot generator.[1]

Design of the optical bar moved rapidly and by 14 February 1967, a Concept Review of the unit was presented and accepted with certain reservations.[2] It was felt that increased confidence in the reliability of the optical coatings was required and that the thermal effect of the optical bar caging mechanism needed further study. Also, the rotary commutator for in-flight instrumentation was not shown on the concept layouts. There was also a potential overweight problem. At the time the design was proposed (21 July 1966) the weight of the optical bars was estimated at 1096 pounds. The design presented at the Concept Review was estimated at 1256 pounds.

The design of the optical bar progressed and on 7 June 1967 the Preliminary Design Review (PDR) was held.[3] Although the general design was accepted, there was some concern about the thermal analyses and approval was not granted until the required data was supplied to the Agency. The estimated weight of the Optical Bars was now 1068 pounds, but some uncertainty still existed in the weight allocations on the Sensor Subsystem.

Soon after the PDR a study program was undertaken to determine system performance with an interchangeable filter. The filter initially proposed was located in back of the field lens group and had to be selected prior to launch. The study showed that an interchangeable filter plate could be located between the second and third elements in the field corrector group without causing serious degradation in the optical design performance. A concept review of the interchangeable filter was held on 20 September 1967 and the new filter location was approved.[4] In addition to having the least effect on optical performance in that location, it was also the most favorable for a mechanical redesign and convenience of installation.

The Optical Bar Critical Design Review (CDR), held on 31 July 1968,

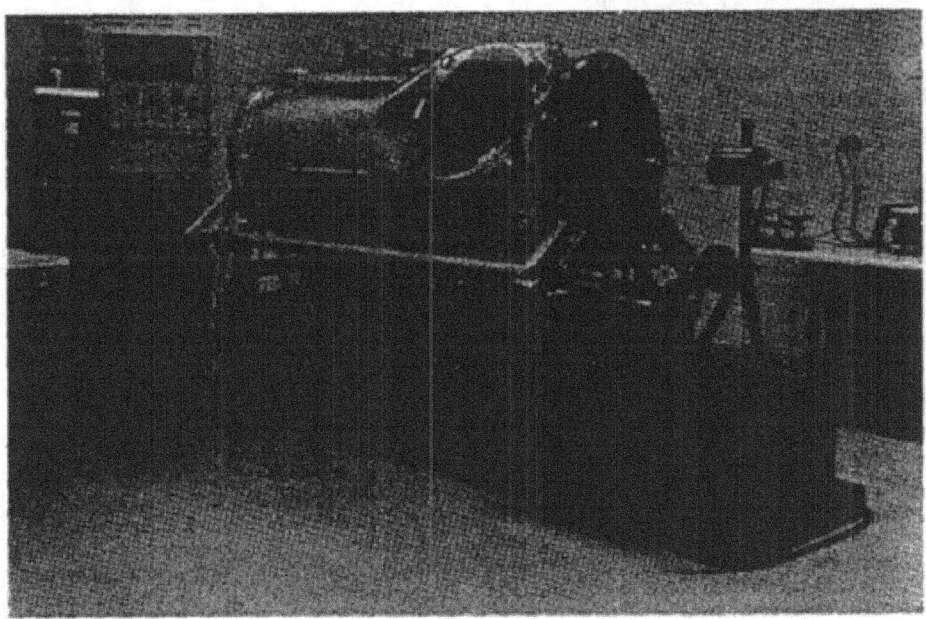

Optical Bar in Dynamic Balancing Machine

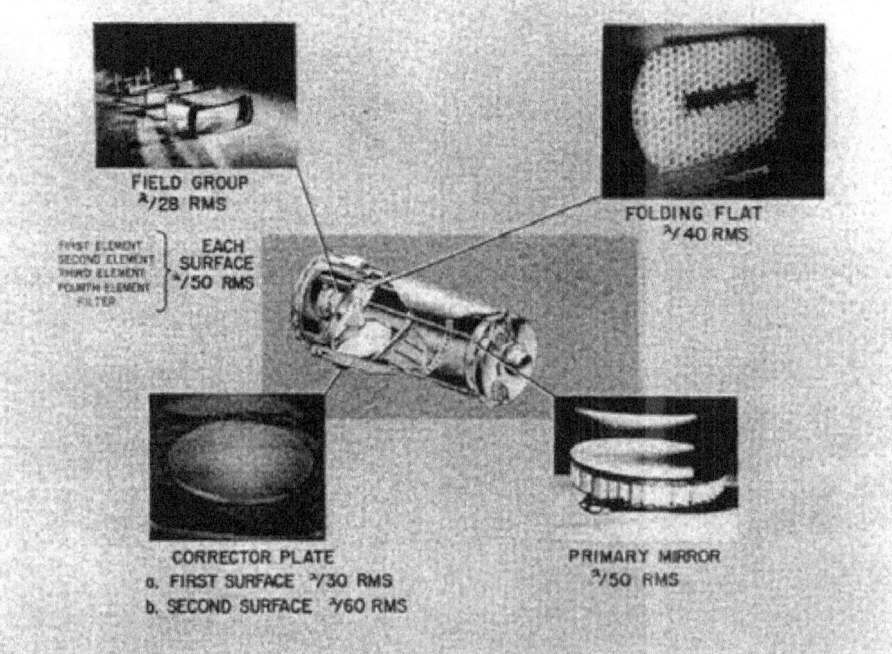

Hexagon Camera Optical Components

opened with a description of the changes in the design since the PDR.[5] A former problem with butt welds on the diagonal mirror support bracket was eliminated by reducing the number of butt welds. An optical bar parking brake was not required since the drag caused by the bearings, slip rings and encoder bearings was sufficient to offset the disturbing torques encountered during launch and orbit.

It was determined that the weight of the optical bars was within budget based on actual hardware weight analyses (1227 pounds). During a discussion of pressurization, the project engineer indicated that a pressure seal could be implemented on the Development Model, but that a slightly different design would be required on the Engineering Model. (The rotating optical bar seal developed by P. Pressel was a true state-of-the-art development). After a review of the CDR data package, the Agency granted approval to begin fabrication of the optical bar.

Camera Support Frame Assembly

The Frame Assembly supports the optical bars, loopers, and associated film path components. The frame controls the dynamic loads transmitted to the optical bars and serves to position the optical bars relative to the Attitude Control System (ACS) in orbit. It must maintain positioning accuracy of the camera optics through launch environment, thermal distortion of the vehicle in orbit, and during operation of the ACS thrusters.

The initial frame drawings were released to the subcontractor (█████████) on 26 January 1967 to enable them to meet the completion dates of the Mass and Thermal Model frames.

The Frame Concept Review, held on 24 February 1967, revealed no significant design problems and was approved.[1] However, additional structural analyses were required to determine the effects of cantilevering the looper assembly on the

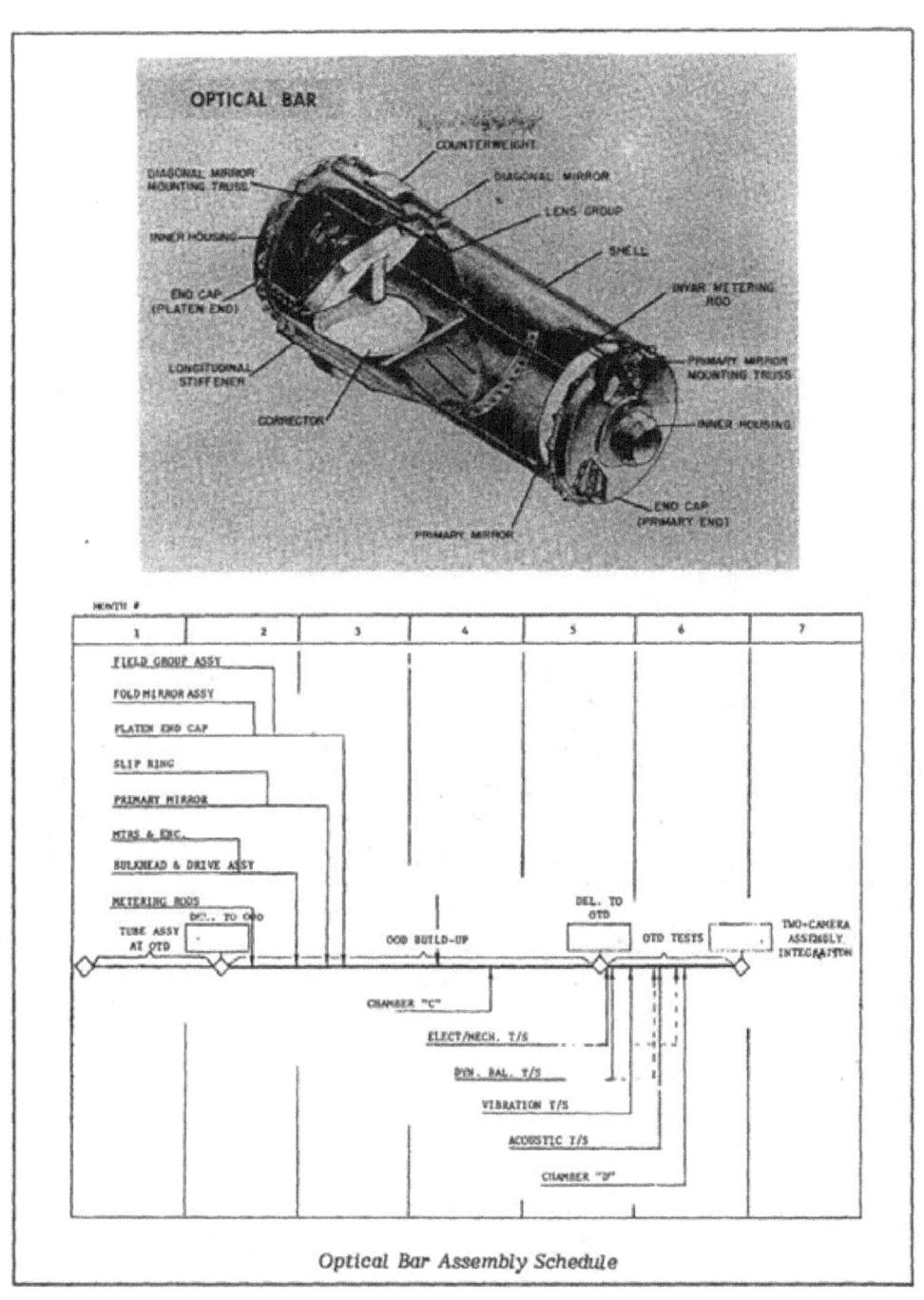

Optical Bar Assembly Schedule

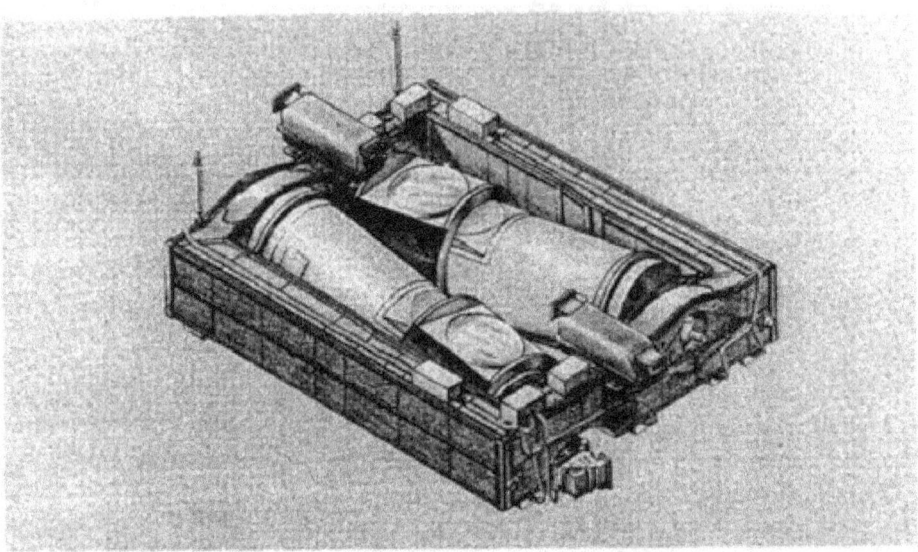

Two Camera Assembly (TCA) Typical of SV-1 through SV-16 Configuration

frame. Studies were also needed to determine the impact that a change in the stereo angle would have on the frame weight and performance.

The estimated weight of the frame listed in Perkin-Elmer's proposal was 280 pounds. By January 1967, the weight estimate increased to 383 pounds and edged upward to 388 pounds at the Concept Review. By 15 June 1967, a decrease in the weight of the optical bars permitted a decrease in the frame weight to 339 pounds.

The Frame PDR (23 May 1967) was only the second PDR held on the program since the award of contract.[2] Don Patterson, CIA Sensor Subsystem Program Director, took this opportunity to make a few comments on the nature of a Preliminary Design Review. He remarked that the PDR is a review for the customer during which time the contractor attempts to convince the customer that he has done the job needed to prove out the design of the functional unit being reviewed. The review should be supported by all the documentation resulting from engineering studies, analyses, reviews, experiments and tests that have been performed to confirm that the design will meet the performance requirements. This effort should satisfy all concerned that the design will perform within the functional specifications governing it and so can be approved for detail drawing preparation.

Patterson also defined SETS' (customer technical consultant) role in the PDR as adviser to the customer regarding the adequacy of the information presented to make a technical evaluation of the design, and in determining whether the design actually meets the requirements of the specifications.

The day that the Frame PDR was held, a TWX message was sent by the CIA to the NRO advising them that the frame was designed in accordance with Perkin-Elmer's original Hexagon proposal which required the frame to fit within a 90-inch diameter shell.[3] Informal discussions with the Special Project Office managers responsible for the Satellite Vehicle design, however, indicated that the diameter of the Satellite vehicle would be 120 inches regardless of the Sensor Subsystem envelope diameter of 90 inches.

The message went on to state, "At the time of the proposal by the Sensor

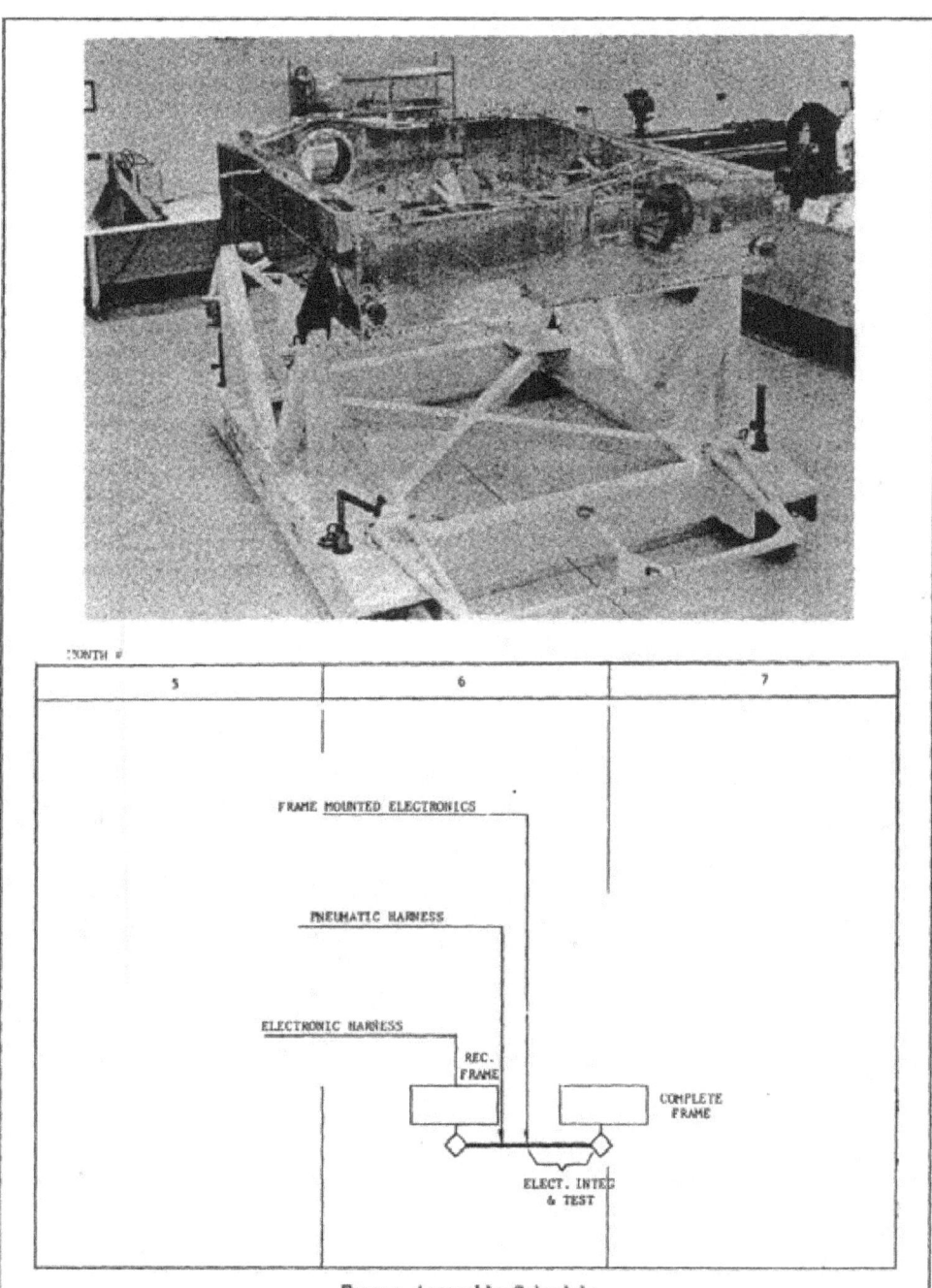

Frame Assembly Schedule

Subsystem contractors, design value was given in rating the proposals for reducing diameter and length. The reduction of the diameter to 90 inches was made with some compromise to weight and simplicity of the frame design. If holding the diameter to 90 inches is not required or necessarily desired, some improvement can be realized by redesigning the frame. With approximately a 6-inch increase in width, a simple box girder structure can be used with a resulting weight decrease of about 27 pounds." Approval for the design change was granted, but not without schedule slippage and increased cost ▓▓▓▓ estimate).[4]

During the Frame PDR, considerable discussion centered on Perkin-Elmer's approach to the frame environment test. The frame design was approved with some reservations noted by Patterson.[5] The customer reserved the right to review the frame design for possible modifications when an "agreed-upon" Performance Specification was approved and when a Perkin-Elmer document specifying the expected Sensor Subsystem environment was prepared.

Design and analysis of the ideal frame corner design was initiated and involved the use of a "box" corner construction. However, problems associated with interface requirements between the Sensor Subsystem and the Satellite Vehicle continued.[6]

The Frame CDR, held 31 July 1967, included a review of the design changes since the Frame PDR.[7] It was reported at this meeting that the frame corners were redesigned to reduce lateral deflection and a strain test performed on a test corner of the new design produced one-fifth the deflection per static load originally expected. The frame at the time of CDR was two pounds under the budget weight (388 pounds). According to the CDR meeting minutes, "both the frame and the frame-mounted electronics are adequately designed from a thermal standpoint." The reliability of the frame was, "in the design and not the numerics." The CDR was subsequently approved by the customer.

On 19 and 20 December 1967, an Interface Working Group meeting was held and covered the total Sensor Subsystem/Satellite Vehicle configuration. The result of this meeting was that the camera frame was lowered 11 inches below the Satellite Vehicle roll axis and the supply reel rotated 90°. This necessitated the lengthening of the fore and aft two-camera assembly envelope and a revision of the Supply Assembly envelope.[8]

Film Drive Assembly

The fine film transport system consists of two functional units; the Platen Assembly and the Film Drive Assembly. The Film Drive Assembly transports the film from the Looper Assembly, to the Platen Assembly, and then back to the Looper Assembly. It supports the twister assembly which permits a loop of film to enter the oscillating Platen Assembly, track precisely through the focal plane, and then exit back into the fixed Film Drive Assembly.

It should be noted that the "air bar twister" is the key component in the "optical bar" configuration. The lack of this device in the configuration devised by Itek was the major drawback in its design. Without the twister, Itek engineers were forced to locate both the film supply spool and the film takeup spool on the rotating optical bar, resulting in a cumbersome design that had little chance of working effectively.

The use of an air bar twister device was first noted in an obscure and unsolicited Perkin-Elmer camera proposal submitted to the government in 1962. The idea lay dormant, however, since the proposal was never converted into a contract. A diagram of the proposed camera is included in the twister patent issued to Perkin-Elmer in 1969.[1]

Charles (Don) Cowles, the inventor of the twister recalled the story of how the idea for the twister developed. "Initially (1955) I worked on a Perkin-Elmer program codenamed Projector Project. Later (1960) Perkin-Elmer was awarded a follow-on to that program. The equipment required air bars and the engineering group I supervised was assigned to develop them."[2]

"All skew bars (air bars) on the

camera system were initially fixed. When we first breadboarded the film transport system, we very quickly learned that it is impossible to align rollers and air bars accurately enough by trial and error. It has to be accomplished by geometrical precision. If you want an air bar to work at an angle of 45° to the roller, you have to measure that angle to seconds of an arc. Also, the rollers must be exactly parallel to within seconds. Errors must be corrected by actual measurement and not by observing the action of the film motion and correcting the position of the air bar and rollers by trial and error. That's a long and bitter lesson that took a lot of adrenaline, late hours, and painful argument to learn."

Don Cowles and his engineers soon discovered that even perfect alignment was not the complete answer. The film strips were not straight and drifted laterally. "We had to go to some degree of self-alignment and developed a pivoted air bar," said Cowles.

"Little did we know what a breakthrough the pivoted air bar was at that time. The pivot point is determined mathematically and if the center of the film tracks exactly over the pivot point, there is no torque applied to the air bar resulting in neutral equilibrium. It's only when the film center moves off the pivot point that there is a moment generated to rotate the bar." Cowles also noted that if a pair of air bars are used, one air bar must be pivoted and the other fixed.

In 1962, Cowles was assigned to proposal activity, primarily in aerial reconnaissance cameras. "We had developed a variety of optical configurations," recalled Cowles, "and were struggling to obtain a high duty cycle from a scanning type panoramic camera. A scanning optical system produces a 50% duty cycle since half the time it scans the earth and during the other half it scans the inside of the camera. In addition, it's necessary to start and stop the film to eliminate waste."

Cowles understood the action and geometry of the pivoted air bar. He realized that when the air bar rotates

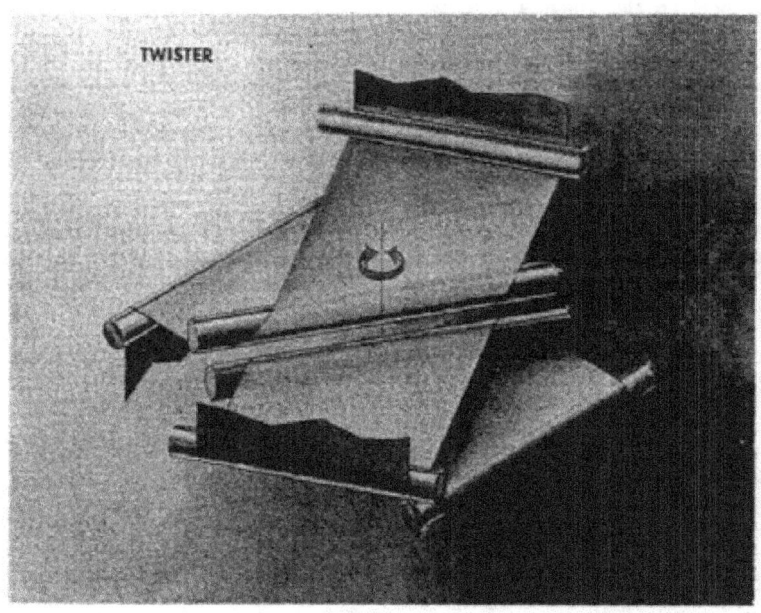

The Air Bar Twister Device

because the outgoing web wants to move laterally while the incoming web does not, there is no net error generated dynamically in the outgoing web. Because of the selection of the pivot point, there is no velocity or lateral position variation caused by the pivoting motion. "Once this is fully understood," said Cowles, "it follows that a pair of air bars positioned on the same pivot point will automatically act to remove the film twist put in by the first air bar by the action of the second air bar without generating any dynamic or tracking errors. I also realized that the twister could rotate a film web 180°." Cowles incorporated the twister device into the camera proposal, but since there was no interest in the camera system, the twister concept lay dormant for the next two years.

In June 1964, Cowles was assigned to work on the CIA parametric study and later on Phase I of the Fulcrum program (September 1964). It was at this time that the need for a twister device once again surfaced.

During one of the conceptual conferences held on the Phase I Fulcrum program, a group of project personnel was discussing the problem of derotating an image of a rotating optical system so that the image could be placed on moving film at the focal plane. Cowles suggested that instead of attempting to use a massive prism to derotate the optical image it might be a good idea to "derotate" the film instead. This of course meant twisting the film a full 120°. "How is this to be accomplished?" someone asked. "With an air bar twister," replied Cowles. "When I dropped that idea for consideration," said Cowles, "it was just like a bombshell. We were seeking some way of combining a continuously rotating optical system with an oscillating platen and the solution was filed away in an obscure proposal somewhere in the company archives."

To prove the feasibility of the twister concept, a design layout was started on 12 October 1964 and a breadboard was constructed and tested.[3] It was an unqualified success. As expressed in the breadboard test report, "The behavior of the twister mechanism was without fault during static and dynamic tests with film transport.[4] It may be successfully employed in any system which requires the direction of film travel in a single plane to be changed during transport, or during periods when the film is stationary. It accomplishes its function passively without introducing path-length changes and without disturbing alignment."

Earle Brown, a staff engineer on the Fulcrum program, went through a process of elimination in the selection of a system configuration for the Phase I Fulcrum program.[5] A point was finally reached where a decision had to be made between an arrangement called "Turnstile 1" and "Turnstile 2." The first arrangement, Turnstile 1 required a twister, Turnstile 2 did not. It was decided by Project Management that the Turnstile 2 configuration would be recommended to the customer (CIA). A full scale breadboard model of the system (called the "Cocktail Shaker") was constructed and tested. As it turned out, the customer had little interest in Perkin-Elmer's proposal and the "Cocktail Shaker" was eventually dismantled. The kindest of critics called it "interesting," others called it an "abomination."

The twister device was once again relegated to the back shelf until March 1965, when Perkin-Elmer was asked by the CIA to continue the "optical bar" design started by Itek. Studies revealed that the only way the "optical bar" configuration could be made to work effectively was by the incorporation of the air bar twister. The operation of the twister was verified in subsequent tests and included in the Sensor Subsystem design.[6]

Prior to the Film Drive Concept Review, the proposed method of using an input and an output steerer was studied and it was decided that one steerer could be used as a bidirectional steerer mechanism, thereby eliminating one steerer. The Film Drive design presented at the Concept Review, held on 2 March 1967, included two drive capstans, a data chamber, a film marking device (hole punch), a twister, and associated structure. The estimated weight of the design presented was 17 pounds (two more pounds then the design in

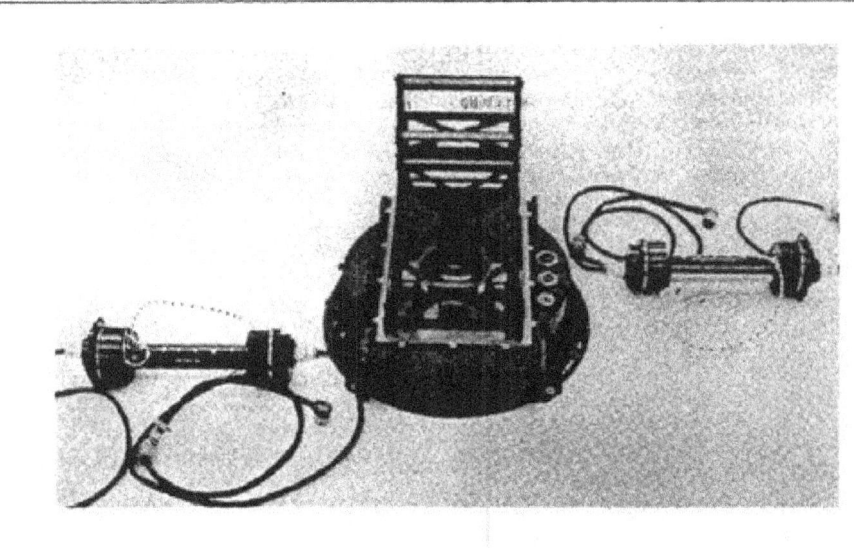

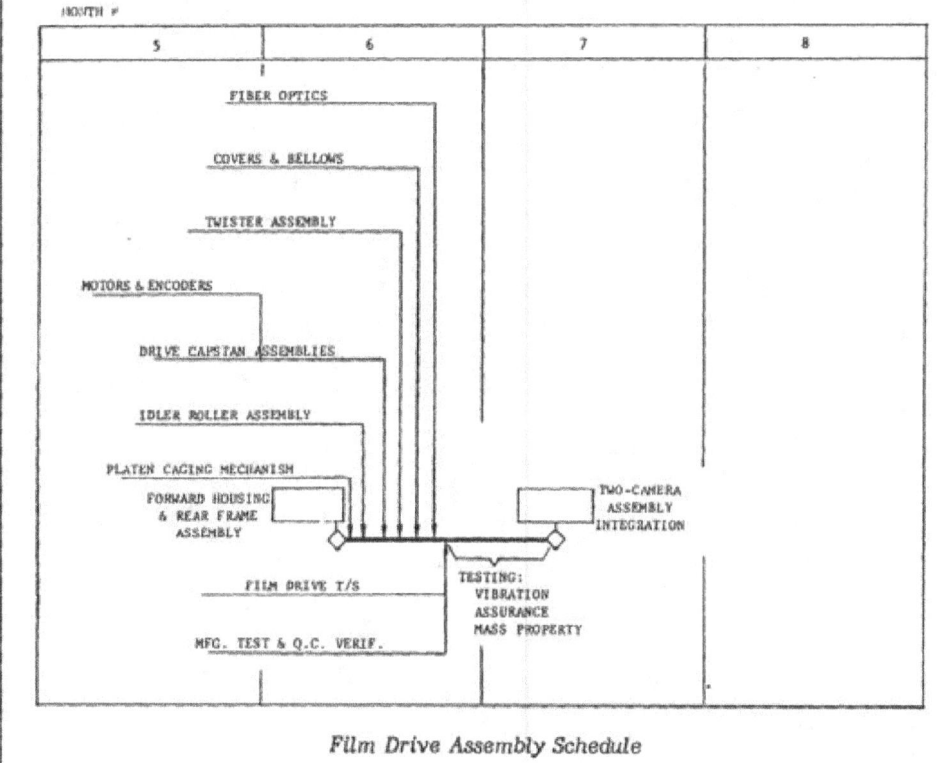

Film Drive Assembly Schedule

the Hexagon proposal).

The Film Drive Concept Review was approved as presented, however, the Agency directed the project engineer to concentrate his attention on the following: measure breakaway force needed to move film if it assumes a permanent set, investigate the possibility of using a hole punch mechanism with higher reliability, determine the effect of short scan operation on the location of the hole punch, investigate the possibility of increasing film travel time prior to exposure, and determine if a problem exists in flexing the control cabling at the low temperatures encountered during the mission.[7]

In May 1967, the Agency instructed the project engineer to proceed with design studies for a film marking device (later to be called end-of-run marker). Thirteen concepts were under consideration at the time. However, maximum effort was to be expended on concepts not relying on a "punch" device since failure of this type of echniques and an approach which limited the variety of parts that could be used on the system."

The efforts of all the functions in a reliability department are directed to the production of a reliable system. These efforts are reflected in a reliability number (from zero to one) which is used as a measuring tool. "However, we didn't make it a numbers game," said Karachuk. "We emphasized the design support alysis backing up the decision to remove the steerer."

On 28 August 1968, the Critical Design Review was presented and approved by the customer.[9] The weight of the unit (18½ pounds) was well within the budget established at that time.

The "punch" device for marking the film was abandoned in favor of a non-contact optical method.

Platen Assembly

The Platen Assembly is perhaps the most critical functional unit in the Sensor Subsystem since it must accurately position the film in the focal plane as the image is exposed on its surface.

Design effort on the Platen Assembly began in the summer of 1965. Since it was "buried" in one end of the Optical Bar Assembly, the biggest problem faced by the design engineers assigned to this task was packaging the many platen components and subassemblies in a small volume of space. The Platen design went through several versions and modifications prior to the design submitted in the Hexagon proposal in July 1966.

Soon after the award of contract, a workshop session was held (2 November 1966) to discuss the platen image motion compensation servo problem and establish a course of action leading to its solution.[1] It was decided that SETS was to continue pursuing a cam drive design and Perkin-Elmer was to investigate the use of gas bearings, devise a system to isolate the platen from the main bearings during film exposure, and build and test a servo breadboard consisting of a simulated platen mounted in commercial grade bearings, with an appropriate torquer, feedback transducers and electronics.

On 6 January 1967, a concept review of the platen bearing arrangement was held to determine which of three mechanical arrangements presented at the meeting would be adopted. The final decision was to locate the platen bearing in the Optical Bar Assembly and react the torquer against ground (frame structure).[2]

Investigation of gas bearings for application in the Platen Assembly was terminated in January 1967 because of the reliability factors involved and the need for large amounts of gas. It was also decided that the torques produced by the large diameter ball bearings mounted in the Optical Bar Assembly were well within the capability of the platen servo.[3]

A Preliminary Platen Concept Review was held on 13 January 1967.[4] The purpose of the meeting was to update the customer on Perkin-Elmer's progress on the Platen and review SETS' activity on the cam drive.

In the initial stages of the Platen design, an effort was made to avoid the use of beryllium. It was soon apparent that this would not be possible. The estimated weight of the Platen Assembly listed in the

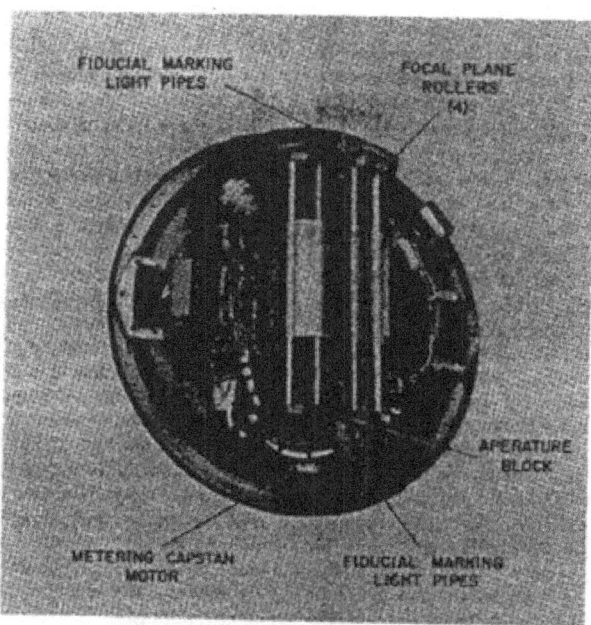

Front View of Platen Assembly

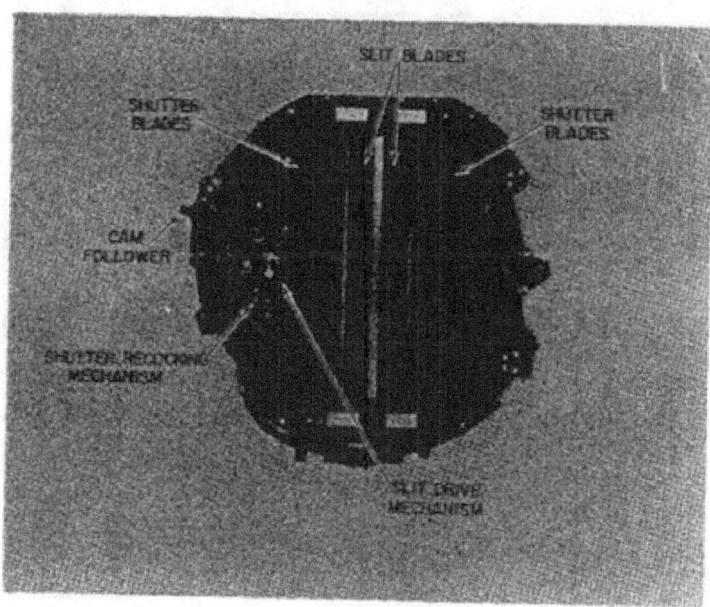

Slit and Shutter Assembly

Hexagon proposal was 10 pounds. By February 1967, the weight had increased to 32.9 pounds, and during a technical review held in March 1967, it was revealed that the proposal weight was unrealistic and that a weight reduction program would be initiated to lower the Platen weight which at that point in time had creeped up to 37.7 pounds.[5] The weight of the platen was eventually reduced to 35 pounds.

On 5 May 1967, a Platen Assembly Concept Review was presented and approved.[6,7] The three concerns voiced by the customer at that time was lubrication of the platen bearings, the availability of the platen servo test results, and the effects that pressurization of the Sensor Subsystem would have on the platen.

In June 1967, a meeting was held to review the two methods of obtaining image motion compensation in the Sensor Subsystem. Both Perkin-Elmer's and the SETS' approach were discussed. The Agency decided to go with the Perkin-Elmer servo design and terminate SETS cam approach.[8]

As the Platen design progressed and breadboards were constructed to test the feasibility of the designs, problems began to surface. Each was carefully analyzed and the mechanisms were either modified or completely redesigned. For example, in August 1967, test results of the shutter produced consistent failures. Regardless of the corrective measures employed, the failures continued. A new design was undertaken which was superior to the original design.[9]

On 7 February 1968, the Platen Assembly Preliminary Design Review was held and the final design approved. Platen drawings were released for the Engineering and Development Models in April 1968.[10] Seven months later, the Platen Assembly Critical Design Review was presented to the customer, and after the completion of some action items imposed by the customer, Perkin-Elmer received approval to release the drawings to the production department.[11]

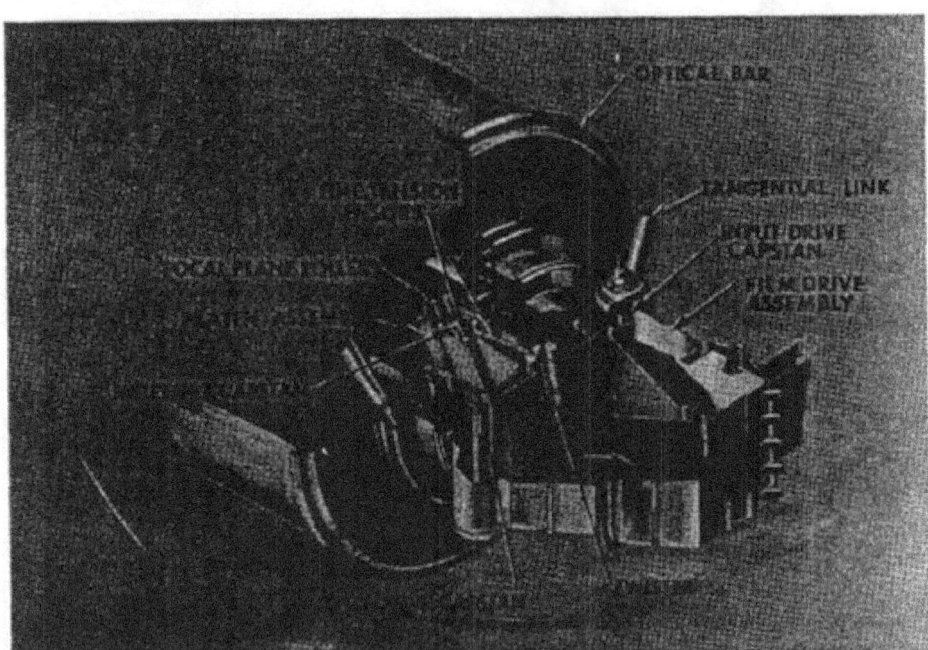

Platen Assembly Components

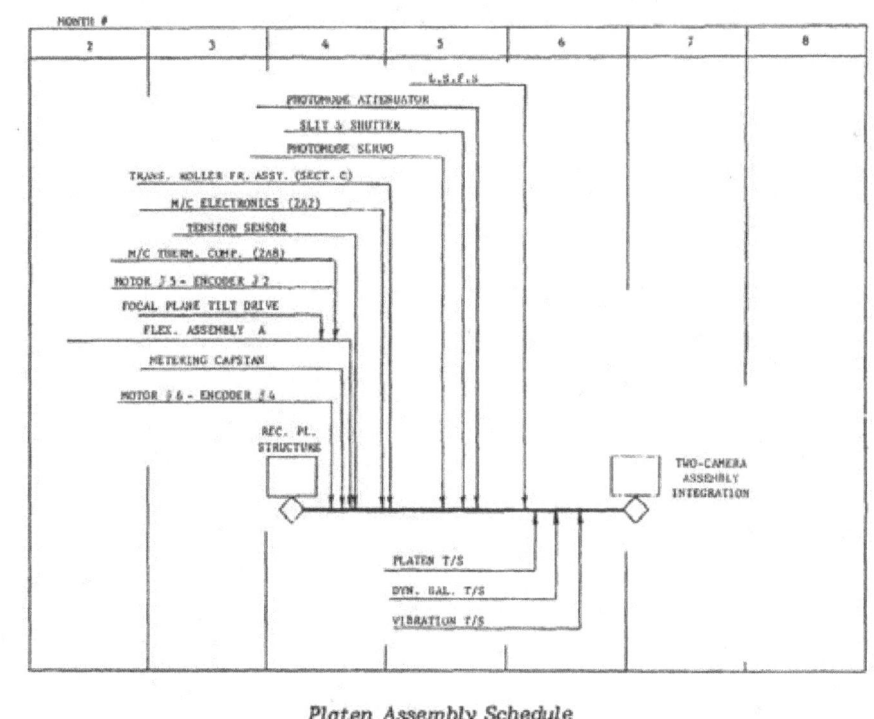

Platen Assembly Schedule

Supply Assembly

The coarse film transport system includes all functional units that operate at a nominally constant speed during photography and recycle. These include the Supply Assembly, the Looper Assembly, the Film Path Components, and the Take-Up Assembly. Film is unwound from the Supply Assembly at a constant rate, transported through one side of the Looper Assembly to the fine film transport system (Film Drive and Platen Assemblies), returned to the other side of the Looper Assembly, and transported at a constant rate to the Take-Up Assembly.

The Supply Assembly supports, protects, and drives the film supply for both the forward-looking camera and the aft-looking camera. Each supply reel carries 104,000 feet of 6.6-inch wide Type 1414 film and weighs 890 pounds. The major components of the Supply Assembly are: core assembly, primary support structure, motor brake, caging assembly, pressure enclosure, film exit vestibule, and servo electronics.

Early in the Sensor Subsystem design development phase (May 1965), the axis of the Supply Assembly was parallel to the launch vehicle roll axis. This configuration was presented in Perkin-Elmer's Hexagon proposal (July 1966).

At the time of contract award, a Supply Assembly design layout existed based on a 2000-pound film load and information in the Sensor Subsystem Performance Specification Book.[1] A revised performance specification was received on 24 October 1966 which initiated a change in the design concept of the Supply Assembly. This change caused a considerable weight increase.[2] A study was conducted to determine the impact of the new specification on the proposed Supply Assembly design. Of the nine design concepts developed, an arrangement with the spool axis parallel to the launch vehicle pitch axis was recommended.[3] Lack of confirmation on film density and thickness led to confusion in the calculation of the overall dimensions of the Supply Assembly.[4]

Design work continued on the Supply Assembly and on 2 March 1967 a Concept Review was held. However, because of an overweight condition, the Concept Review was not approved. The estimated weight of the Supply Asembly in the Hexagon proposal was 500 pounds. The design presented at the Concept Review was estimated at 846 pounds. A weight reduction study was started and revealed that very little weight could be eliminated unless the flanges were omitted. This left only two choices for a weight decrease of 200 pounds; the first arrangement was with the spool axis parallel to the roll axis of the vehicle, the second with the spool axis parallel to the pitch axis of the vehicle.[5] The study showed the latter to be the lighter design.

The next Concept Review on the Supply Assembly was held on 27 April 1967. In the configuration presented, the Supply Assembly axis was parallel to the vehicle roll axis. An alternate approach was also presented with the supply axis aligned to the vehicle pitch axis. This design was not considered acceptable because it projected into space reserved for the Orbiting Command module in the aft section of the vehicle.[6] The Concept Review was approved and the design continued. By 15 May 1967 the Supply Assembly weight was reduced to 740 pounds.

In August 1967, in response to a request from the customer, Perkin-Elmer submitted a proposal for a Supply Assembly oriented with its axis of rotation parallel to the vehicle pitch axis.[7,8]

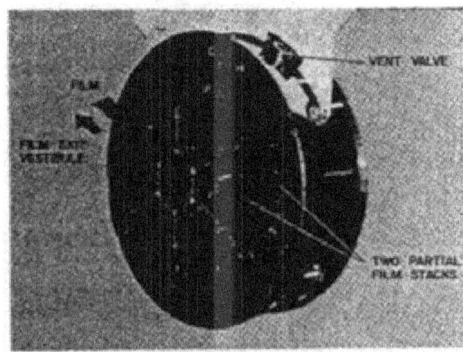

Film Supply Assembly

In December 1967, a customer decision was made to reorient the reference supply design with the spool axis parallel to the vehicle pitch axis. Although this change did not affect the overall design of the Supply Assembly developed to this point, engineering time had to be spent analyzing changes to accommodate the supply reorientation (e.g., film exit locations, loading changes, etc.). The allowable weight at this point was 770 pounds and the estimated weight was 798 pounds.

A year later, the Supply Assembly PDR was presented (21 February 1968). Considerable time was spent discussing the effects of film splices on the edge sensors, the selection of the structural material, the relationship of the film winding procedure to the core design, and the pressure-venting-condensation relationship in both ascent and normal operating conditions. At the conclusion of the review it was agreed that areas requiring particular emphasis in the next phase of design were the caging arrangement, the overweight problem, the core design, the condensation problem, and the over-budget reliability problem. The allowable weight of the Supply Assembly at this time was 823 pounds and the estimated weight was 851 pounds.[9]

The next five months were spent conducting tests and preparing drawings for the Critical Design Review. On 20 June 1968, the 50-inch diameter film stack underwent an acceleration test. It was first subjected to a 1.0g acceleration for a duration of 120 seconds to check out the operation of the centrifuge, the instruments, and the recording equipment. The film stack was then subjected to both 4.5g test and an 8.0g test, both for a duration of 120 seconds. Some residual deflection occurred at the 8.0g test.[10]

The Supply Assembly CDR was finally held on 25 November 1968.[11] The project engineer opened the CDR with a discussion of the areas redesigned since the PDR and then reviewed the functional specifications of the Supply Assembly. Inconsistencies between the functional and the design specifications were immediately pointed out by ▆▆▆▆▆▆ (customer representative). The areas of concern were the rewind stacking and film tension specifications. ▆▆▆▆▆▆ directed the recording secretary to document this as "highly irregular" and then recessed the CDR for a caucus with customer representatives attending the review to decide whether to continue the meeting. ▆▆▆▆▆▆ acting as spokesman, reconvened the CDR and requested that the meeting continue as outlined by the agenda contingent upon agreement that copies of the fully approved Functional and Design Specifications were delivered to the customer by 4 December 1968. The project engineer of the Supply Assembly agreed to this requirement.

The CDR continued without further problems and after responding to a number of action items, Perkin-Elmer received approval for the Supply CDR. The estimated weight at CDR was 865.7 pounds and the allowable weight was 860 pounds.

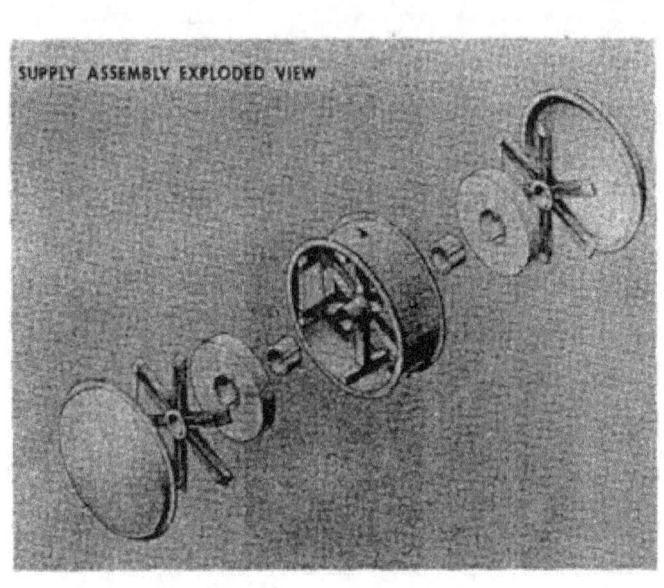

SUPPLY ASSEMBLY EXPLODED VIEW

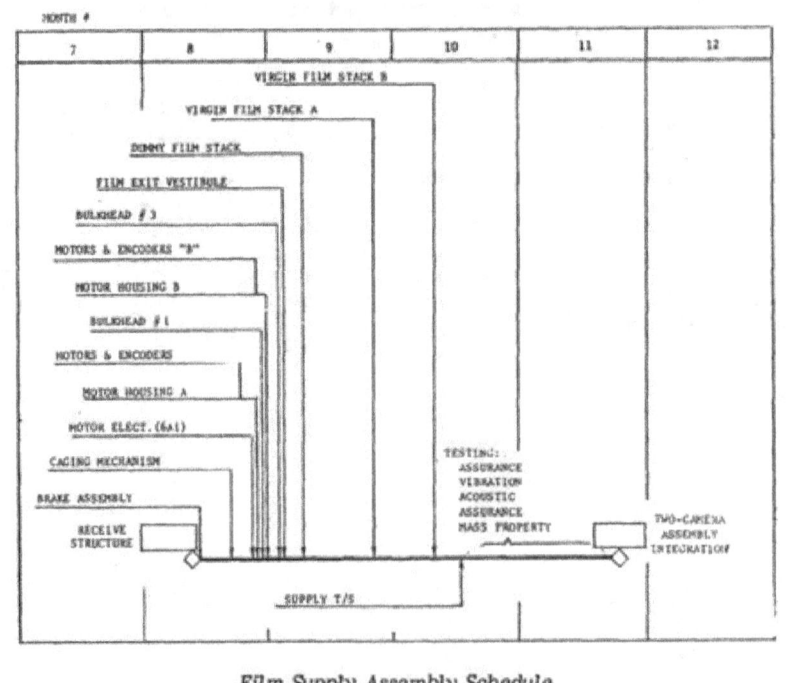

Film Supply Assembly Schedule

Looper Assembly

The function of the Looper Assembly is to store sufficient film before and after exposure to accommodate the intermittent film motion in the Platen Assembly in combination with the continuous film motion at the Supply and Takeup Assemblies. The Looper Assembly consists of three basic subassemblies: the frame and mounting structure, the carriage, and the tension sensor. An initial configuration of the Sensor Subsystem in a May 1965 study shows the Looper Assembly positioned behind the Optical Bar Assembly. As the design progressed, the Looper Assemblies were moved to the Frame Support Assembly.

The layouts and breadboard tests started before the award of contract were presented at a Looper Assembly Concept Review held on 23 February 1967. However, the review was regarded as a "preliminary" concept review because the functional specifications, the control system requirements, and the structural analysis were not sufficiently defined or complete.[1]

The Looper Assembly Concept Review was rescheduled to 4 May 1967. The design and supporting documentation presented at this review was more detailed and the Concept Review was approved with certain reservations.[2,3]

Throughout its design, the Looper Assembly was affected by changes in other parts of the Sensor Subsystem. The film path configuration determined to a large extent the position of the Looper Assemblies on the Frame Assembly.

Progress on the Looper Assembly continued and a PDR was scheduled. However, it was delayed by Perkin-Elmer because fundamental questions existed on the basic requirements. A letter from C. W. Besserer, Manager of the SETS organization, contained a critique of the Looper PDR Design Package submitted to the customer on 13 July 1967.[4] It was unfavorable and mentioned that if the PDR had not been delayed by Perkin-Elmer, SETS would have recommended the delay because the PDR Design Package was inadequate.

The Looper PDR was rescheduled to 26 September 1967 and once again was reviewed by SETS. Their response left little doubt of SETS' disapproval of the quality of engineering work on the Looper design. A major portion of the Besserer memo discussed the philosophy of formal design reviews and the requirements of a properly prepared design review package supporting a PDR.[5]

Regardless of SETS' dissatisfaction with the Looper PDR Package, the PDR was held on the scheduled date.[6] At the conclusion of the Looper PDR, ███████, customer representative, noted that the initial concern about the readiness for a Looper PDR expressed by SETS was precipitated by an evaluation of the PDR package and that the Looper PDR presentation answered most of the questions. He suggested that future PDR Design Packages contain as much of the available material as possible rather than reserving it for the PDR presentation.

In a letter dated a few days later, Don Patterson supported this view and made suggestions which would prevent the repetition of the confusion caused by an incomplete PDR Design Package.[7] Approval of the Looper PDR was deferred by Don Patterson until a system trade-off could be accomplished based on new information regarding additional frames of photography required for cycling the film through the forward-looking camera. In addition, he listed items that represented areas of concern. These included the need for additional analysis of the launch survival of the coarse tension sensor, roller design and alignment, the need for a more complete thermal analysis, and the addition of performance requirements in the Looper Design Specification. On 15 November 1967, K. W. Patrick, General Manager of the Optical Technology Division, forwarded a technical report which completed Perkin-Elmer's response to all of Patterson's requirements.[8] Final approval of the Looper PDR was received on 28 November 1967.[9]

Detailed design of the Looper Assembly continued in preparation for the CDR. However, the Looper design was impacted by additional changes outside of

Looper Assembly

its envelope. Rearrangement of the film path and the film steerer components and modifications of the Frame Support Assembly created problems for the Looper designers. An added system requirement for short scan photography resulted in a looper carriage imbalance. The Looper cover required numerous changes because of fabrication difficulties and a change to a pressurized film path. In July 1967, the looper designers had to respond to yet another change due to the replacement of the air-bar steerer by an articulated steerer design. Incorporation of the new steerer design required a change to the looper structure, the upright supports, and the cover.

These changes, however, improved the overall design and reliability of the Looper Assembly and by the time the Looper CDR was presented on 18 November 1968, the unit was generally simplified.[10] Although the customer approved the CDR, some concerns were expressed by ▓▓▓▓▓▓ who felt that there were limited tests on the Looper design and that the differences between the Engineering Model and Development Model Looper designs were significant. He believed that this would undermine confidence in the Looper performance until the Development Model Looper was tested. There was also concern that the reliability estimate did not meet budget allocations. The total weight of the Looper Assemblies at CDR was 71 pounds. The proposal estimate of July 1966 was 40 pounds. It must be pointed out that in addition to carrying components not included in the proposal estimate, looper requirement changes were also responsible for a large portion of the increased weight.

Looper Assembly (SV-1 through SV-16 Configuration)

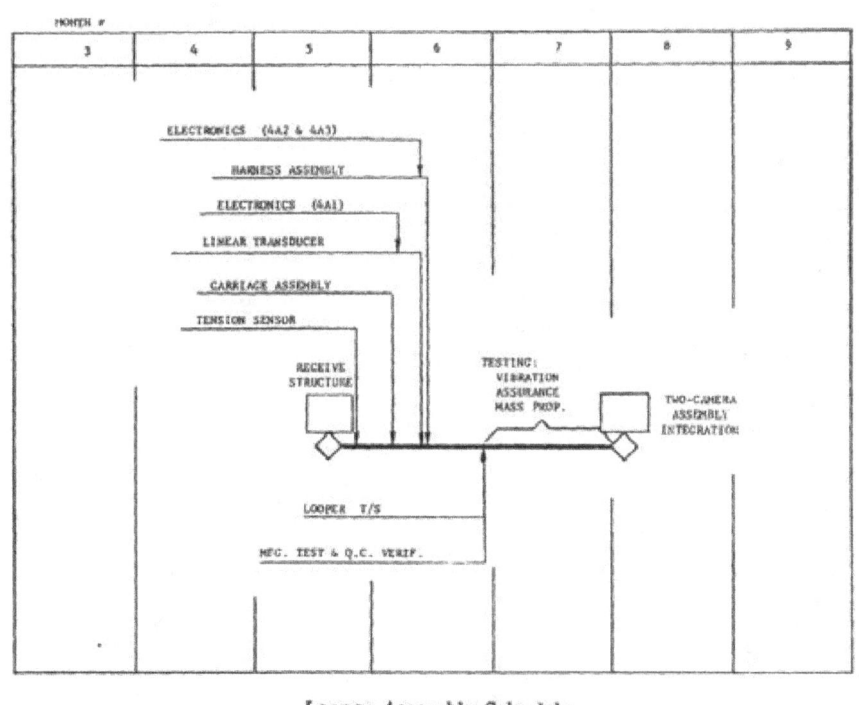

Looper Assembly Schedule

Film Path Assemblies

The Film Path Assemblies perform two functions; they provide guidance and support to the film as it is transported between major functional assemblies, and a protective enclosure to prevent film exposure, damage, or unacceptable environmental conditions. The basic components of the Film Path Assemblies are air bars, rollers, and steerer mechanisms.

Air Bars

The most critical element in the Sensor Subsystem is the air bar. Development of this important element began in 1960 when Dr. Roderic M. Scott, Chief Engineer of the Reconnaissance Branch in the Electro-Optical Division, suggested the use of air bars for supporting film in an ongoing program. He was aware of the use of air bars in the paper industry in the production of paper rolls, and also by the Eastman Kodak Company to transport wet film in the coating process. However, these applications used a great amount of air at high pressures to lift the transported material.

Charles D. Cowles, supervisor of the mechanical group on one of the programs managed by Dr. Scott, was assigned to develop the air bar. Cowles, and a few of the engineers in his group, began to experiment with what were initially called "air rollers" in an application requiring low pressure and a limited amount of air supply. The group tried various approaches with little success.

Dr. Robert E. Hufnagel, a staff engineer, was asked by Dr. Scott to spend some time on the air bar problem and recommend a new approach. During a visit to the laboratory, which at that time was in the Connecticut Avenue plant in Norwalk, Dr. Hufnagel observed the air bar experiments. Hufnagel recalled that, "They were simply round bars with some holes drilled into them. During the tests, an attempt was made to transport film over the air rollers but the film edges scraped on the air roller surface and the air rollers just couldn't lift the film without a horrendous amount of air flow.

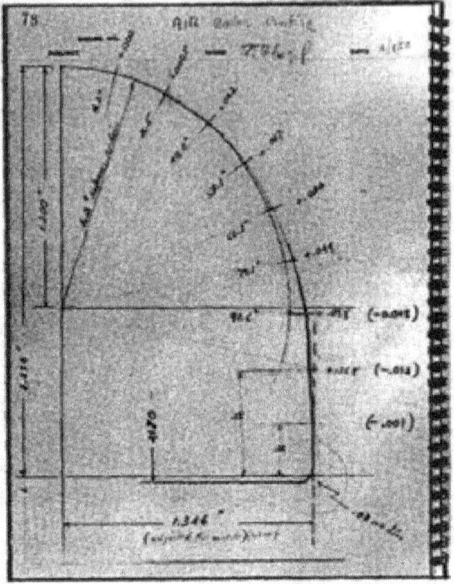

Original Computations of Air Bar Design from Dr. Robert E. Hufnagel's Engineering Notebook

In fact," commented Hufnagel, "I'm not sure if the engineers ever got the round air rollers to lift the film at lower pressures."[1]

The air roller was a great idea — except it wasn't working," said Hufnagel. "I gave it some thought and my first entry in my engineering notebook on 4 April 1960 was a basic equation for fluid flow. It was clear that the equations were too complicated and that I would not be successful with that approach. Later on that same day, I recorded some general thoughts on the essence of a solution to the air roller problem. At that time I was not even addressing the "skew" problem but just trying to transport film over an air roller, or "air bar" as I chose to call it."[2]

By 6 April, Hufnagel had developed the profile of the air bar and determined the radius of curvature, the thickness of the air cushion (0.0012-inch), and the air flow (0.95 ft^3 per minute). The unusual profile (D-shaped) and the porous material (sintered) of which it was first fabricated, made it difficult to produce, except at high rejection rates. The original theory was

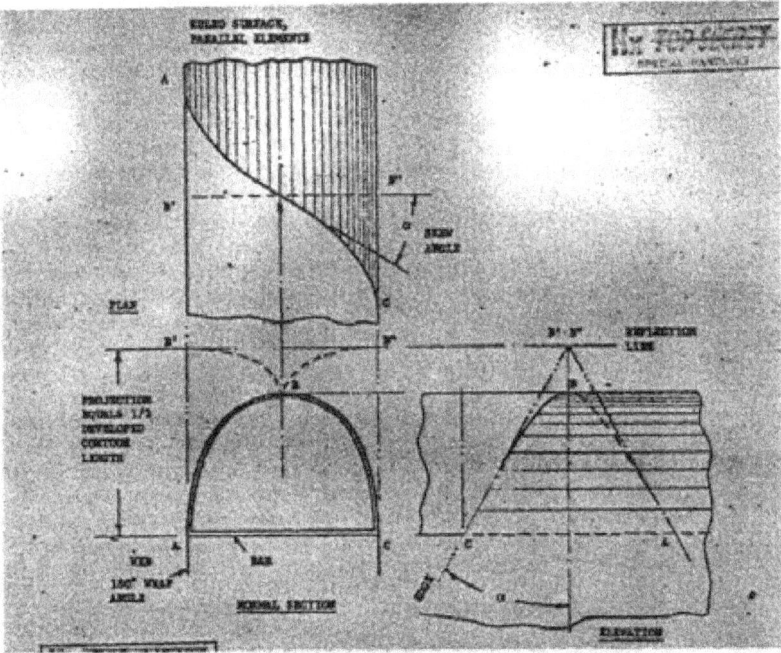

Conceptual Drawing of Air Bar Profile

based on uniform porosity of the air bar and not a hole pattern. By 5 May 1960, Hufnagel and the engineers soon realized that it didn't make any difference. The most important factor was the air bar contour and that the radius of curvature of the air bar surface continually increased as it approached the edge.

The orifice size selected was a 0.005-inch hole based on the minimum size which could be drilled by conventional methods. Hole patterns were varied and extra holes were added during development testing to achieve adequate lift under the specified conditions.

When trouble was experienced in producing holes to the required tolerances, inserts were designed which could be precision machined and cemented into counterbored cavities at the hole locations.

The air bar design fulfilled the requirements of the program. Air bars were used in place of rollers in applications where the arrangement of the film transport caused the film to take a helical path around the air bar, thereby accommodating any lateral motion without friction.

When Perkin-Elmer entered Phase I of the Fulcrum program, it was necessary to produce a full size model of what was called the "Cocktail Shaker." Cowles was assigned to develop the film transport for the system and used air bars in several locations of the film path that required "skewing" or film directional changes. This activity provided additional experience in air bar production and use.

In support of the proposal for Phase I (January 1965), air bar tests were conducted to determine the characteristics of the air bar design. The total system consumption was 895 psi of nitrogen in a 2 ft^3 volume at a pressure of 1.75-inch of mercury in the air bars. It was noted that a small change in any of the parameters (flow rate, supply pressure, ambient pressure, etc.) caused the film to either drag on the air bar or vibrate violently. At best, the air bar was a marginal component that worked within a very narrow set of conditions.[3,4]

During the period in which Perkin-Elmer was investigating the advantages and disadvantages of the F' and M' systems (July 1965), Cowles wrote a one-page technical report forecasting the gas consumption for the air bars in both systems. In it, he referenced the breadboard tests conducted during the preparation of the Phase I Fulcrum proposal submitted to the customer in January 1965. The characteristics of a "sample bar" were listed.[5]

The importance of the air bar was emphasized in Perkin-Elmer's Hexagon proposal (July 1966), which stated that exhaustive experimental and development work was carried on to develop the air bar on a previous program and that the component was "thoroughly proven." While it is true that the air bar worked successfully on a previous program, it did so under a specific set of conditions (low, constant speed with no reversals or start and stops). The application of an air bar to the requirements of the Hexagon Sensor Subsystem proved to be a most difficult and perplexing problem.

Soon after the award of the Hexagon contract (February 1967) a Concept Review was held to approve the general configuration of the air bar.[6] As a result of this meeting, tests were planned to accomplish the following: continue analysis to verify the optimum contour, analyze flow rate for film flotation, determine flow rate at specific velocities, and analyze structural and vibration characteristics of the air bar.[7] A test in October 1967 determined the flow rates (under static conditions) for typical air bars of the skew, twister, and steerer types.[8]

Dynamic air bar performance, to that point in time, was evaluated only by observing the operation of steerers and

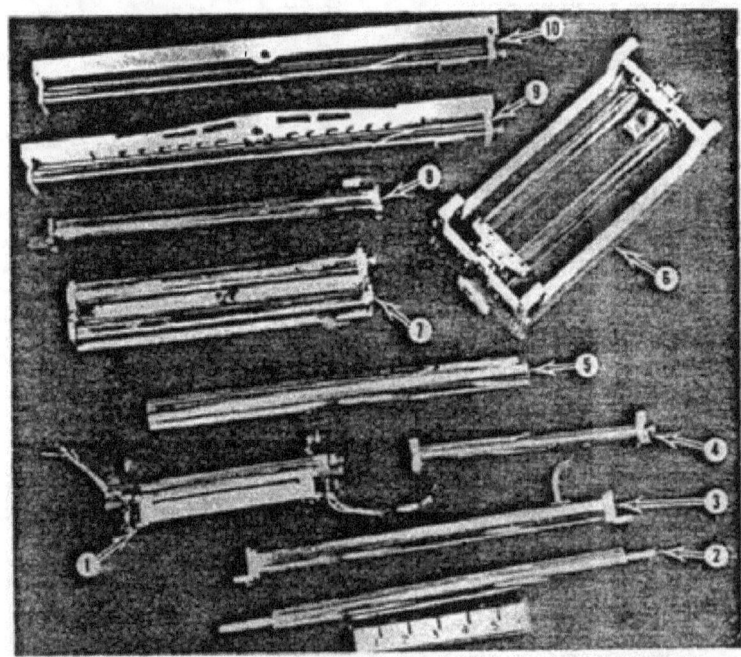

1. Capacitance Edge Guide Sensor; 2. & 3. Spew-Type Air Bar; 4. In-Line Air Bar Assembly; 5. Twister Air Bar; 6. Experimental Test Fixture; 7. Twister Air Bars; 8. Skew Air Bar; 9. & 10. Self-Aligning Air Bar Assemblies

other assemblies which used air bars. What was needed was a special apparatus to measure film lift at numerous points over the air bars. This device could demonstrate the effects of changing the number and/or pattern of holes. It could determine the uniformity of film lift under all conditions of film speed, tension, film type, gas pressure, etc.).[9]

In December 1967, the new air bar tests were completed.[10] The experiment was designed to measure the relative force required to transport film over two air bars (steerer type) under conditions of maximum film tension and various nitrogen pressures. Under certain conditions of operation, emulsion buildup (or scabs) developed on the air bars. It was known prior to the tests that scabs were frequently discovered on air bars used in film transport experiments on the test beds. Further tests were recommended to isolate the cause of the "scabbing."

The design of the air bar continued based on the latest test results. It was decided to eliminate the inserts and drill the 0.005-inch diameter hole directly in the air bar extrusion.[11] A test to determine the coefficient of friction of the air bar surface was conducted in April 1968 and film lift measurements were made in May 1968 resulting in a slight modification of the air bar design.[12,13] But the problem of "scabbing" air bars did not go away. In July 1968, the Film Drive project engineer reported that tests were hampered by occasional emulsion buildup on the steerer air bars.[14]

The problem of emulsion buildup on the air bars occurred more frequently and finally summed up in the August 1968 Monthly Technical Report.[15] "Emulsion buildup on air bars has been a continuing problem on the test bed steerer air bars and the cause for it on these bars has not yet been isolated. Emulsion buildup also occurred on the twister air bars during the initial tests; however, it seems to have been alleviated by correcting the angle of approach of the film on the air bar (from negative to positive angle)."

Meanwhile SETS, disturbed by the many reports of air bar problems, recommended that analysis and modeling of the air bar be attempted to more clearly understand the characteristics of air bar operation.[16]

Twister Assembly tests on the film path simulator and the abbreviated film path began to highlight a new air bar problem. Local drying of the film in the region of each of the air bar holes caused furrow-like corrugations on the length of the film.[17]

By January 1969, it was becoming apparent that a serious problem was developing on one of the most critical components in the Sensor Subsystem and that a systematic test to completely understand the operation of the air bar was absolutely mandatory. A development test plan for the air bar was written 17 January 1969.[18]

In 21 March 1969, an attempt was made by the Perkin-Elmer System Engineering Group to obtain a qualitative picture of air bar operation. A flux plotting technique was used to simulate flow patterns for several air bars.[19] SETS, in the meantime, issued a final report on Air Bar Mathematical Modeling (May 1969).[20] Although both of these analyses were interesting, they had little practical value in solving the air bar problem.

The new Perkin-Elmer air bar development program (started in January 1969) began to show results in June 1969. The use of teflon coatings and the addition of grooves held a great deal of promise.[21,22] The tests verified a basic fact of air bar operation — the edges of the film will always contact the air bar surface regardless of the hole pattern.[23]

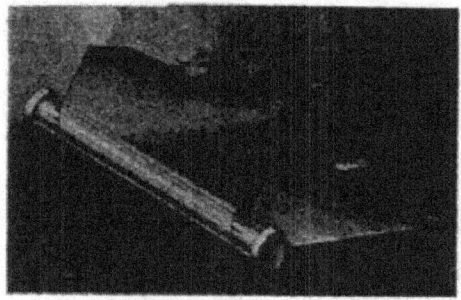

Sketch of Improved Air Bar Design

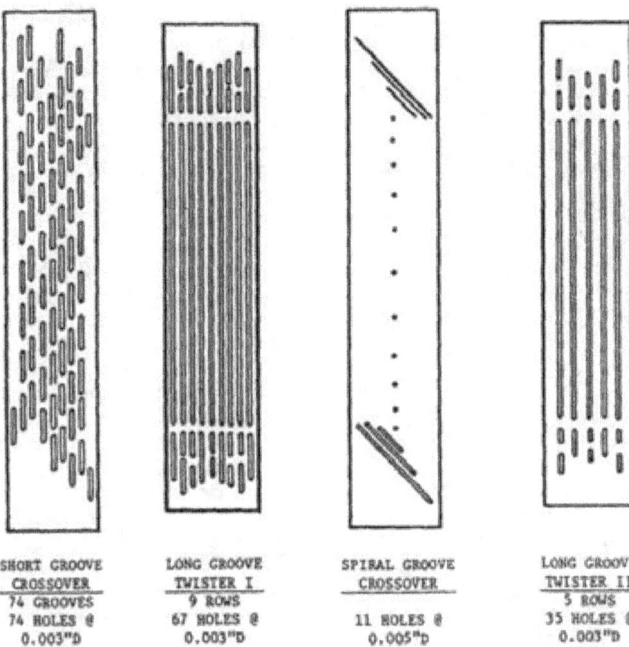

Various Air Bar Groove Patterns Developed During Experimental Tests

By 1970, the air bar problem was resolved. Tests proved that the air bars would work effectively if a filter was used to prevent hole clogging and if hole and groove patterns were selected for each particular application of the air bar. The grooved patterns reduced wrinkling caused by impingement of the dry gas on the film and provided a more uniform gas support under film. The use of a teflon coating was, of course, mandatory to prevent emulsion buildup. Because of inherent film curl, the edges of the film could not be lifted unless air pressure and supply requirements were exceeded.

In retrospect, it appears that the air bar problem was too complex to be solved mathematically. It required over two and one-half years of engineering effort to design a workable air bar for Hexagon Sensor Subsystem mission requirements. A solution was finally achieved by concentrated testing and observation using flight-designed air bars.

The lesson to be learned from the "air-bar" story is not to permit a component to jeopardize the success of a major program by accepting its use in the equipment without adequate analysis and testing under the required operating conditions.

Rollers

A second component of great importance to the Sensor Subsystem film transport is the roller. A reference to rollers was made in a supplementary report written in support of the Phase I Fulcrum proposal in January 1965.[24] "Conventional bends of the film web are made with precision lightweight rollers with integral low friction bearings. Each roller will be mounted with its axis precisely normal to the direction of web travel so that there is pure rolling contact."

Although this critical element was used in 186 locations (SV-1 configuration) throughout the Sensor Subsystem (94 rollers in the "A" film path and 92 in the "B" path), little was noted in the Perkin-Elmer Hexagon proposal (July 1966) relating to roller design. Perhaps it was assumed that this element would create no difficulty since Perkin-Elmer had successfully designed rollers for other programs.

Prior to award of the contract to Perkin-Elmer, a project memorandum noted the progress that had been made in this area (roller design).[25] "As a result of the many inputs to the program, there now exist design layouts and drawings of a large variety of rollers. It is strongly recommended that a study be made of the various designs to establish a minimum number of types and sizes to be used in the final equipment."

Among the many factors to be considered were material (beryllium suggested), mounting method (inclusion of a labyrinth seal to inhibit dirt and lubricant evaporation), and roller size and shape. Apparatus which was designed and built at that time to measure roller characteristics, however, was cannabalized to expedite construction of the Film Path Simulator and had not yet been replaced.

By February 1967, a "standard" roller assembly design was developed for the Hexagon Sensor Subsystem.[26,27] A diameter of 0.600-inch was selected for all rollers in the film path with the exception of the 0.370-inch diameter rollers used at the focal plane which had special requirements (angle of wrap, higher precision, limited space, etc.).

The 0.600-inch roller diameter provided a safety factor above the minimum bending radius of UTB and STB film. The roller and bearing designs met the requirements of minimum inertia, minimum friction, an internal labyrinth seal for lubrication retention and dirt exclusion. The roller assemblies were divided into three classes (bearing precision) and several categories based on application (width, load, etc.).

Studies were initiated to determine the frictional effects of various protective finishes and breadboard testing was conducted to select a lubricant meeting system operation requirements.[28]

A standard mounting for the roller assembly was developed by July 1967 and included a wave spring washer and an oil reservoir (pad) to reduce lubricant migration and outgassing and serve to provide additional protection from film particles.[29]

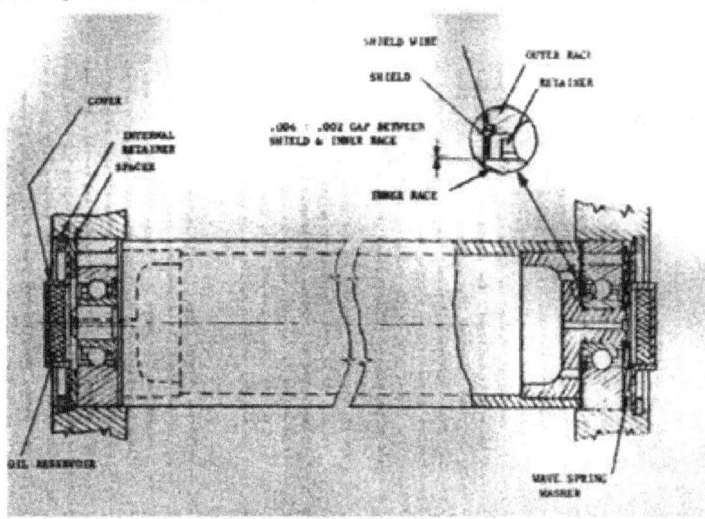

"Standard" Film Roller Design

Roller development test plans were completed in September 1967 and outlined ten separate tasks to verify roller properties and design.[30] A month later a partial shipment of beryllium rollers was received.[31] As a result of the information obtained from completed roller tests, vendor consultation, and system servo analysis, the roller design was modified. The bearing outer ring was reduced from 0.625-inch diameter to 0.375-inch to increase the width to outside diameter ratio thereby reducing the possibility of bearing "hang-up" during launch vibration.[32]

Roller drawings were approved and released for production by February 1968. Tests on the rollers, however, continued and roller assembly characteristics for eight different film path applications were determined.[33] Completed roller assemblies were incorporated into various breadboards and simulators by June 1968 — and then it happened.[34] The first roller failure occurred on the Film Path Simulator, followed by four roller failures on the Abbreviated Film Path Simulator. Stub shafts on the roller assemblies were snapping off. The failed rollers were examined to determine whether the failures were caused by impact or fatigue and it was soon discovered that impact loading was the cause of all the failures.[35]

Several factors had contributed in lowering the impact properties of the beryllium rollers. These included highly stressed material, surface porosity, surface defects caused by machining, and poorly machined undercuts on the stub shafts. Measures were taken to correct these deficiencies.

A data and application list released in March 1969 identifies ten roller designs.[36] By selecting the proper combination of roller and stub shaft material (lockalloy, stainless steel, aluminum, or beryllium) it is now possible to design a roller assembly tailored to meet any requirement on the Sensor Subsystem.

Steerers

The third film path component critical to the operation of the film transport system is the steerer mechanism. Previous to the Hexagon Sensor Subsystem, cameras developed at Perkin-Elmer relied on passive film transport systems requiring only the precise alignment of film supporting rollers. Charles D. Cowles, the inventor of the "twister", recalled the transition to active steering systems.[37] "As a result of our experience on a previous program, we learned that there is a limit as to how far you can go without active steering. We thought we might be able to use a passive steering system on Phase I of the Fulcrum program, but I was never convinced that we could. We lucked out with a passive film transport system on the previous program, and we did have our problems, but it was as far as we could go

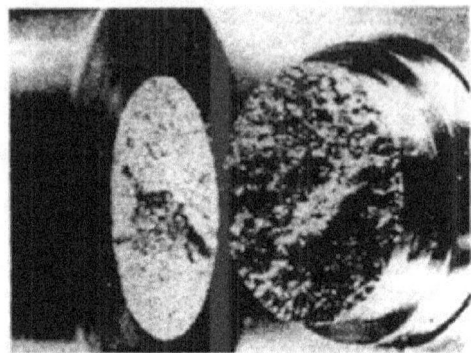

End View of Beryllium Shaft Fracture (10X)

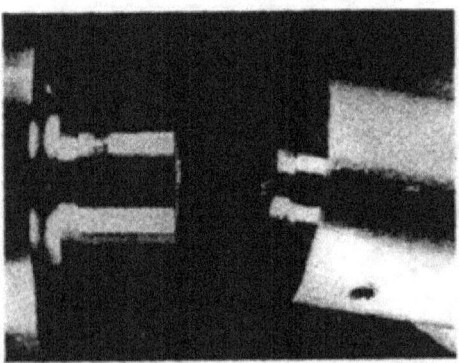

Side View of Beryllium Shaft Fracture (3.5X)

without active steering."

As Perkin-Elmer engineers began working on Phase I of the Fulcrum program, it became apparent that active steering would be required. The Phase I fulcrum proposal (22 January 1965) included a reference to film transport guidance.[38] "To provide for accurate placement of photographic data relative to the width of the film, and to increase the control of IMC, a closed loop edge guidance system is located at the entrance to the shuttle platen. The lateral displacement means required for steering and shuttle motion for IMC is ideally furnished by simple displacement of the skew (air) bar which feeds the shuttle."

Although initial reports and presentations published in May and June 1965 do not refer to any steering devices for film guidance, a sketch presented in a meeting in September 1965 shows the location of "steerers" at the entrance to the platen and the re-entry vehicles.[39,40]

A Design Review Package sent to the CIA by Perkin-Elmer (12 November 1965) contained a reference to film transport devices.[41] "Control components which center the film web without resort to flanges anywhere in the system are included. Active or passive control is necessary only to provide long-term lateral stability of position. Lateral drift rate will be adequately limited by designing to take advantage of the smoothing effect of rollers."

In a customer presentation, held on 9 December 1965, a viewgraph was shown referring to an experiment on steerers.[42] The objective was to develop a passive steering system using "Lorig" aligners, crown rollers, and pivoted roller devices.

Throughout 1965, Perkin-Elmer engineers vacillated between the use of air bars (using a closed loop edge guidance system) and a passive roller arrangement that made use of shaped rollers or self-aligning pivoted rollers. But by May 1966, the emphasis seemed to be on passive steering devices.

A project memorandum, dated 13 May 1966, described experiments on tapered, conical, and pivoted roller arrangements for film path guidance.[43] An analysis of the geometry of the crowned roller was reported in a memorandum dated 9 June 1966.[44] A few weeks prior to the submittal of the Perkin-Elmer Hexagon proposal in July 1966, a project memorandum describing film steering cylindrical rollers was published on 16 June 1966.[45]

The steerer approach finally recommended in the Perkin-Elmer Hexagon proposal, however, was based on the use of air bars and edge guidance sensors.[46] The proposal dismissed the use of film guidance rollers. "If flanges, crowned rollers, or other brute force methods are used to control tracking, film can be damaged and system performance affected resulting in loss of information."

Investigation of active (air bar) steering devices began soon after the award of contract (10 October 1966).[47] However, the proponents of the film steerer devices using pivoted rollers were still undaunted.[48] The first Sensor Subsystem Monthly Technical Report refers to this activity.[49] "Concurrently with the work on the servo controlled (air bar) steering units, a low priority effort has been put on the development of a purely mechanical (self-energizing steerer roller) device." Although this was the last reference in the monthly reports on steerer rollers, the advocates of this approach once again tried to stimulate interest in a passive steerer approach.[50] "The film guidance philosophy expressed in the proposal (Hexagon) and presently being used, requires high precision of manufacture and alignment of the entire system. This memo recommends a relaxation of this philosophy which could well be applicable to the major part of the system and permit considerable relaxation of manufacturing and alignment tolerances. It is further recommended that active steps be taken to explore the possibilities of employing guidance devices which do not require air bars." The memorandum continued with a discussion of crowned rollers, free-pivoted rollers, and controlled pivoted rollers.

A report written by the customer's consultant (SETS) on 23 May 1967 discusses an analysis which was performed to without active steering."

As Perkin-Elmer engineers began working on Phase I of the Fulcrum program, it became apparent that active steering would be required. The Phase I fulcrum proposal (22 January 1965) included a reference to film transport guidance.[38] "To provide for accurate placement of photographic data relative to the width of the film, and to increase the control of IMC, a closed loop edge guidance system is located at the entrance to the shuttle platen. The lateral displacement means required for steering and shuttle motion for IMC is ideally furnished by simple displacement of the skew (air) bar which feeds the shuttle."

Although initial reports and presentations published in May and June 1965 do not refer to any steering devices for film guidance, a sketch presented in a meeting in September 1965 shows the location of "steerers" at the entrance to the platen and the re-entry vehicles.[39,40]

A Design Review Package sent to the CIA by Perkin-Elmer (12 November 1965) contained a reference to film transport devices.[41] "Control components which center the film web without resort to flanges anywhere in the system are included. Active or passive control is necessary only to provide long-term lateral stability of position. Lateral drift rate will be adequately limited by designing to take advantage of the smoothing effect of rollers."

In a customer presentation, held on 9 December 1965, a viewgraph was shown referring to an experiment on steerers.[42] The objective was to develop a passive steering system using "Lorig" aligners, crown rollers, and pivoted roller devices.

Throughout 1965, Perkin-Elmer engineers vacillated between the use of air bars (using a closed loop edge guidance system) and a passive roller arrangement that made use of shaped rollers or self-aligning pivoted rollers. But by May 1966, the emphasis seemed to be on passive steering devices.

A project memorandum, dated 13 May 1966, described experiments on tapered, conical, and pivoted roller arrangements for film path guidance.[43] An analysis of the geometry of the crowned roller was reported in a memorandum dated 9 June 1966.[44] A few weeks prior to the submittal of the Perkin-Elmer Hexagon proposal in July 1966, a project memorandum describing film steering cylindrical rollers was published on 16 June 1966.[45]

The steerer approach finally recommended in the Perkin-Elmer Hexagon proposal, however, was based on the use of air bars and edge guidance sensors.[46] The proposal dismissed the use of film guidance rollers. "If flanges, crowned rollers, or other brute force methods are used to control tracking, film can be damaged and system performance affected resulting in loss of information."

Investigation of active (air bar) steering devices began soon after the award of contract (10 October 1966).[47] However, the proponents of the film steerer devices using pivoted rollers were still undaunted.[48] The first Sensor Subsystem Monthly Technical Report refers to this activity.[49] "Concurrently with the work on the servo controlled (air bar) steering units, a low priority effort has been put on the development of a purely mechanical (self-energizing steerer roller) device." Although this was the last reference in the monthly reports on steerer rollers, the advocates of this approach once again tried to stimulate interest in a passive steerer approach.[50] "The film guidance philosophy expressed in the proposal (Hexagon) and presently being used, requires high precision of manufacture and alignment of the entire system. This memo recommends a relaxation of this philosophy which could well be applicable to the major part of the system and permit considerable relaxation of manufacturing and alignment tolerances. It is further recommended that active steps be taken to explore the possibilities of employing guidance devices which do not require air bars." The memorandum continued with a discussion of crowned rollers, free-pivoted rollers, and controlled pivoted rollers.

A report written by the customer's consultant (SETS) on 23 May 1967 discusses an analysis which was performed to

Initial Breadboard Steerer and Edge Sensor on Test Bed (Without Film)

determine the guidance capability of contoured rollers and an evaluation of the feasibility of their application to the (Hexagon Sensor Subsystem) film path.[51] "The results indicate that the guidance capability attainable by this technique is too small to be of significant value in reducing the tolerance required on the system." It made no mention of the possible damage that could be caused to the film by shaped rollers.

Development tests and design of the "air bar" steerer continued throughout 1967 and the first part of 1968. A Concept Review of the active steering mechanism (air bar) was held on 8 March 1967. The design was approved and mathematical analysis and breadboarding of the active steerer mechanism continued. Experimental work on the Film Transport Test Bed and the Film Path Simulator was expanded to include the testing of improved steerer designs.

A Film Path Concept Review held on 12 May 1967, recommending active steerers both in the Sensor Subsystem and the forward section of the satellite containing the re-entry vehicles, was approved. By August 1967, problems of instability in the steerable air bar began to surface.[53] Modifications were made to the steerer design on an attempt to solve the problem. Sufficient progress was made to warrant the incorporation of the active air bar steerers into the Sensor Subsystem.[54]

One persistent steerer problem continued throughout steerer testing on the Film Path Simulator. A low-frequency oscillation of the film between the rollers and the air bar occurred whenever the pivoted air bar was disturbed. The oscillation was caused by the sidewards bending of the film between the rollers and the air bar. William A. Newell, a staff engineer on the program, was assigned to the problem. He published his results in January 1968 verifying the measured results by analysis.[55]

A Film Path PDR, on 20 December 1967, reported the progress on the steering tests and presented various engineering analyses supporting the design. The design was approved.[56] However, by April 1968, stability tests, coupled with computer

Film Transport Test Bed

analysis, revealed that the steerer mechanism had resonances at two frequency ranges. Modifications were made in an attempt to solve this problem.[57]

It was now quite apparent that the air bar had problems that could not be corrected by simple redesigns. The basic concept was at fault. While design and test engineers were working feverishly to solve the problems of the "air bar" steerer, the originator of the "air bar" steerer, Charles D. Cowles, and another staff engineer (Walter McCammond) began to discuss the possibilities of using a system of "articulation".

Cowles, of course, had an excellent understanding of film transport systems using rollers. He was aware of "neutral axis twisting" (the neutral axis of the film is defined as the axis along the center of the film length). He understood that neutral axis twisting can be accomplished if the rollers at either end of the film web lie in parallel planes and the misalignment rotation is about the neutral axis of the film only. Cowles applied this theory to a system of passive steering he called "articulation". Cowles and McCammond began working up a design, and using cardboard and paper, they fashioned a mockup of an articulator. They tried to promote their idea as an alternate to the air bar but were unheard.

In the meantime, program management assigned William A. Newell to study the problems of the "air bar" steerer. Newell, just prior to this new assignment, shared an office with Cowles and McCammond and was aware of their discussions on "articulated" film paths.

Newell examined the kinematics of the air bar steerer and other steerers previously proposed and strongly recommended the replacement of the air bar steerer with passive and "steerable" articulators.[58] In his introduction to his study report he stated, "Considerable difficulty has been, and still is, experienced in ensuring the reliability of the film tracking properly throughout the

Air Bar Steerer Components on Test Bed

Sensor Subsystem. The difficulty results from the great length of the film path, the large number of rollers and the components along it, the requirement for recycling the film, and the relative motion of various parts of the film path. In a vague way, it has been assumed that a piece of equipment called a steerer would correct for these effects. This has imposed many severe requirements on this device with resulting difficulty in attaining them. It is believed that the function previously assigned to the steerer should be broken up into two parts. The first of these is a form of articulation of the film path to permit misalignment of the various subassemblies along the film path without affecting the tracking of the film. The second is a true steering problem to keep the remaining sideward displacement of the film along the film path within a reasonable amount. When this is done, the articulation in a passive manner provides six degrees of freedom in the displacement of any major subassembly relative to the others with the steerer only required to handle one degree of freedom involving the side motion of the film."

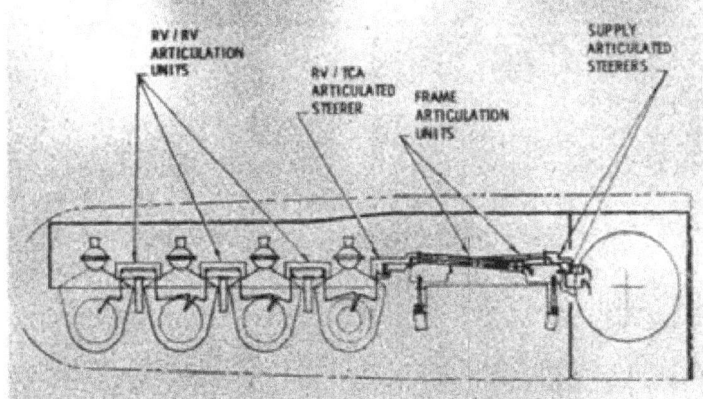
Articulation Units in Sensor Subsystem

Forward Articulator Test Bed

The articulators were implemented into the Sensor Subsystem design and the guidance problem was finally solved. There were minor design and manufacturing problems that had to be overcome, but these were easily corrected.

Although it is true that the steerer problem was solved by the determination of a few, it should be noted that their

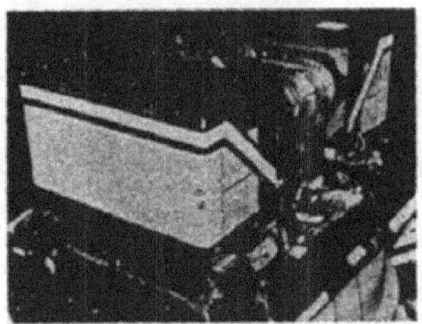

Passive Articulator Qualification Vibration Test

success was based on the efforts of many others who amassed valuable information on roller design, flexure theory, alignment techniques, and numerous tests on the film path test bed and simulator.

Film Path Arrangement

From the time that Perkin-Elmer was asked by the CIA to study the "optical bar" configuration in March 1965, to the award of contract in October 1966, the Sensor Subsystem film path went through several arrangements. At the time the Hexagon contract was awarded to Perkin-Elmer, the satellite vehicle included two re-entry capsules in the forward section, the Sensor Subsystem in the center section, and a supply assembly in the aft section. The space allocation for the Sensor Subsystem

Active Steerer Qualification Vibration Test

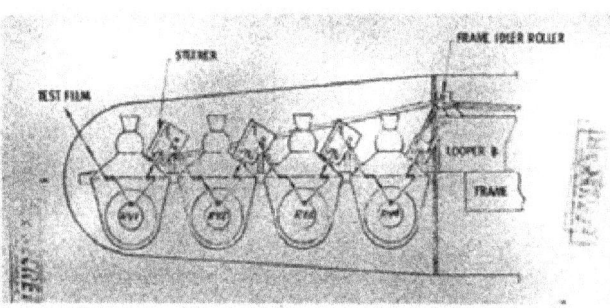

Early Conceptual Drawing of 4-RV Arrangement in Forward Section

was a cylinder 90 inches in diameter and 170 inches in length. Satellite structural information and re-entry capsule (RV) size and configuration at this point in time were very sketchy.

A Film Path Concept Review was held on 19 April 1967 to approve an arrangement for routing the film from the supply assembly, through the Sensor Subsystem, and into the RV's. The film path recommended at this meeting was a side-by-side arrangement of Camera A and Camera B film paths. In the configuration initially recommended in the Hexagon proposal, the film paths were stacked, one above the other. The reason for the change was to reduce weight, decrease the number of air bars in the system, lower the film path and looper assembly center of gravity, and eliminate a cantilevered structure required in the proposal arrangement. The Film Path concept presented at the meeting was approved.[59]

A second concept review of the Film Path arrangement was held on 12 May 1967. Design approaches were presented on air bars, rollers, steerers, film enclosures, film path interconnecting units, and the pneumatic system. The reference design of the film path was a side-by-side arrangement and the number of re-entry vehicles and the orientation of the re-entry vehicles and the supply assembly remained unchanged from the proposal arrangement.[60] This Film Path arrangement was the basis of the Sensor Subsystem design until August 1967, when the customer changed from 2 RV's to 4 RV's in the forward section.

Studies were conducted on various 4-RV arrangements, and exactly one year from the date of contract award, a Forward Film Path Concept Review was held to approve the new 4-RV Forward Film Path arrangement.[61] However, detailed information on the size and configuration of the re-entry vehicle delayed progress.

A Film Path PDR was held on 20 December 1967.[62] The new 4-RV Film Path arrangement was discussed in detail and approved. Just prior to the Film Path PDR, the orientation of the Supply Assembly was changed and the camera frame was lowered.[63] However, there was insufficient time to incorporate this information into the PDR. Progress on the Film Path Assemblies continued, however, definite information on the RV size and shape were still lacking in April 1968.[64]

Final detail drawings were being prepared in July 1968 when a major Film Path redesign occurred. The Film Path guidance system was changed from one based on the air-bar steerer to one using an active articulator. It required several months to assimilate the articulator steerers into the Sensor Subsystem. A Film Path Critical Design Review was held on 19 February 1969, and the new design which had been verified by tests and analysis, was approved.[65] The much needed RV information was received in time to be incorporated into the final Film Path design.

Pneumatic System

The Pneumatic System was considered the responsibility of the Film Path Engineering Group during the initial stages of the Hexagon program. Perkin-Elmer had experience in the design of pneumatic systems on previous programs using air bars. The Perkin-Elmer Protem proposal (May 1965), which was a proposed variation of the Itek "optical bar" design, included air bars which required a Pneumatic Supply.[66] The proposal noted that, "Subcontractors who specialize in the manufacture of gas supply equipment for space applications are available and have complete facilities for production, special testing, and qualification." It appeared that even in the early stages of the Fulcrum program Perkin-Elmer intended to subcontract the Pneumatic Module since it was an unclassified piece of equipment.

Throughout 1965, no mention was made of the Pneumatic System in any program documentation until 30 September 1965 when a presentation was given by Perkin-Elmer to the Associate Contractors on the program.[67] The presentation included the first sketch of the Pneumatic Module and was based on a General Electric Company design submitted to Perkin-Elmer in August 1965.[68] The sizing of a pressure vessel for the Pneumatic System was reported in a technical report that same month.[69] Both a General Electric system layout and the Perkin-Elmer Hexagon proposal showed the Pneumatic Module located in the center section above the rotating optical bars.[70]

Work on the Pneumatic System began immediately after the award of the Hexagon contract and the first Monthly Technical Report contains a complete description of the initial design. The proposed weight of the system was 50 pounds, but by January 1967 it had increased to 150 pounds due to more detailed air supply requirements. Perkin-Elmer and General Electric worked together on developing the initial design and by 21 April 1967, a Concept Review was presented and approved.[71,72] This procurement package for the Pneumatic Supply Module was released to the purchasing department on 26 May 1967. A Perkin-Elmer schematic shows the location of the Pneumatic Module unchanged (above the optical bars.).[73]

Three vendors were asked to bid on the Pneumatic Module; and submitted their proposals by July 1967.[74] In August 1967, the pneumatic distribution lines were changed to accommodate the modification of the Sensor Subsystem from a 2-RV to a 4-RV satellite system. The contract for the Pneumatic Module was scheduled to be awarded by 31 August 1967.[75] However, plans were changed and the was asked to conduct an 8-week study program on 6 November 1967. A meeting was held at that same week to arrange for another quotation and identical action was taken at .[76]

New proposals were received from the three vendors in March 1968. Technical evaluations rated as "good", as "average" and as "poor." The "poor" rating for was mainly due to lack of detailed information and was later upgraded to "good" as a result of a post-proposal meeting. Discussions with the Lockheed company resulted in the relocation of the Pneumatic Module from above the optical bars to the aft bulkhead between the Supply Assembly and the Two-Camera Assembly.[78]

The Pneumatic Module contract was awarded to in April 1968.[79] During contract negotiations with an impasse was reached on 9 May 1968 due to a disagreement on manpower estimates. Discussions with were resumed on 14 May 1968 and the contract was awarded to a few days later.[80,81]

A Pneumatic Module Concept Review was held on 17 July 1968 and approved. A few weeks later (7 August 1968) the PDR was held and accepted.[82] The design continued without any significant problems and by 12 December 1968 all engineering drawings and analysis were completed and a Pneumatic Module CDR was presented and approved.[83] The CDR weight of the Pneumatic Module was 134 pounds (34 pounds of gas included) and was well within the 160-pound allowable weight budget.[84]

Take-Up Assembly

Conceptual design work on the Take-Up Assembly began in the fall of 1965. By 26 April 1966, the first conceptual design report was published. The Take-Up Assembly was designed to carry two spools of 6.6-inch wide film (495 pounds of film on each spool).[1] Four months later, a preliminary design report was issued. Except for a few minor changes and additions it was identical to the April report.[2]

At the time of contract award a Perkin-Elmer design layout of the Take-Up Assembly existed. It was based on a film load of 2000 pounds and the technical concepts described in the reports mentioned above. By 31 October 1966, a new design layout was completed for a 2500-pound film load with the spool axis in the X direction. However, an analysis showed that with the spool axis in the Y direction, the spools provided a much better support for a possible 40g impact. Another advantage of the change to the Y axis was the elimination of skew air bars and all Take-Up Assembly pneumatic lines in the forward section of the satellite vehicle.[3]

A Concept Review was held on 7 January 1967 and approved for design layouts and further analytical studies.[4] The need for a builder roller in the Take-Up Assembly had already been established by previous experimental winding tests. A technical review was held on 8 March 1967 reporting the progress of all auxilliary devices such as the builder roller, builder roller lift-off, peripheral clamp, entry and exit rollers, and cut-and-clamp mechanisms. A four-RV study was also being conducted at this time.[5] The proposal weight of an empty Take-Up Assembly was 178 pounds but had increased to 200 pounds by April 1967.[6] A revised specification with a requirement for 52,000 feet of film per spool necessitated a re-evaluation of the basic dimensions and analyses.[7]

By August 1967, a statement of work for the Take-Up Subsystem was completed and submitted to the Radio Corporation of America. RCA submitted a proposal for the Take-Up Subsystem which encompassed the complete design, development, and fabrication of the Take-Up Assemblies in the 4-RV configuration. RCA maintained a design team on corporate risk funds pending the outcome of the proposal evaluation.[8] The technical review of the RCA proposal states, "In general, the RCA proposal is acceptable both from a technical and implementation standpoint. Technically, RCA shows an awareness of the problems to be solved and proposes an acceptable course of action. From an implementation standpoint, the RCA effort will relieve the critical manpower problem within Perkin-Elmer's Hexagon program team. The basic approach taken by RCA for the design and development of the Take-Up Subsystem is to make maximum possible use of the extensive preliminary design effort already accomplished by Perkin-Elmer."[9] RCA was awarded the contract.

Take-Up Assembly

A Take-Up Subsystem Concept Review was held on 15 November 1967 and approved.[10] The weight of a Take-Up

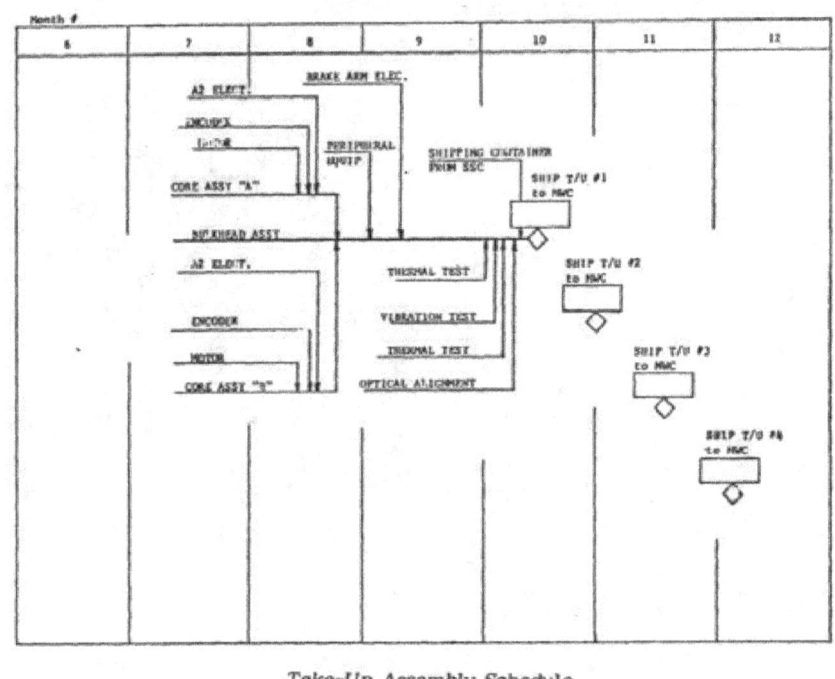

Take-Up Assembly Schedule

Assembly at this stage was estimated to be 153 pounds (allowable weight budget was 165 pounds).[11] Three months later the Take-Up PDR was held and approved.[12] Detailed design work on the Take-Up Subsystem continued without any significant problems and on 12 September 1968, the Take-Up CDR was held and approved. The allowable weight of the Take-Up Assembly was increased to 220 pounds at this point. The weight reported at the CDR was 1/2 pound under.[13]

SYSTEM ELECTRONICS

The development of the electronics system for the Hexagon Sensor Subsystem began soon after Perkin-Elmer was asked by the CIA to continue the design of the "optical bar" configuration started by Itek (March 1965).

Prior to reviewing the evolution of the Hexagon Camera electronics system, it may be useful to study the diagram of the first flight (SV 1) configuration showing the location of the various electronic boxes (RV electronics not included). The electronic boxes are located in three areas: circuits requiring short leads are mounted in the functional units; circuits related to the optical bar, the platen, film drive and active steerer articulators are mounted on the camera frame; the remaining electronic boxes are located in the Supply

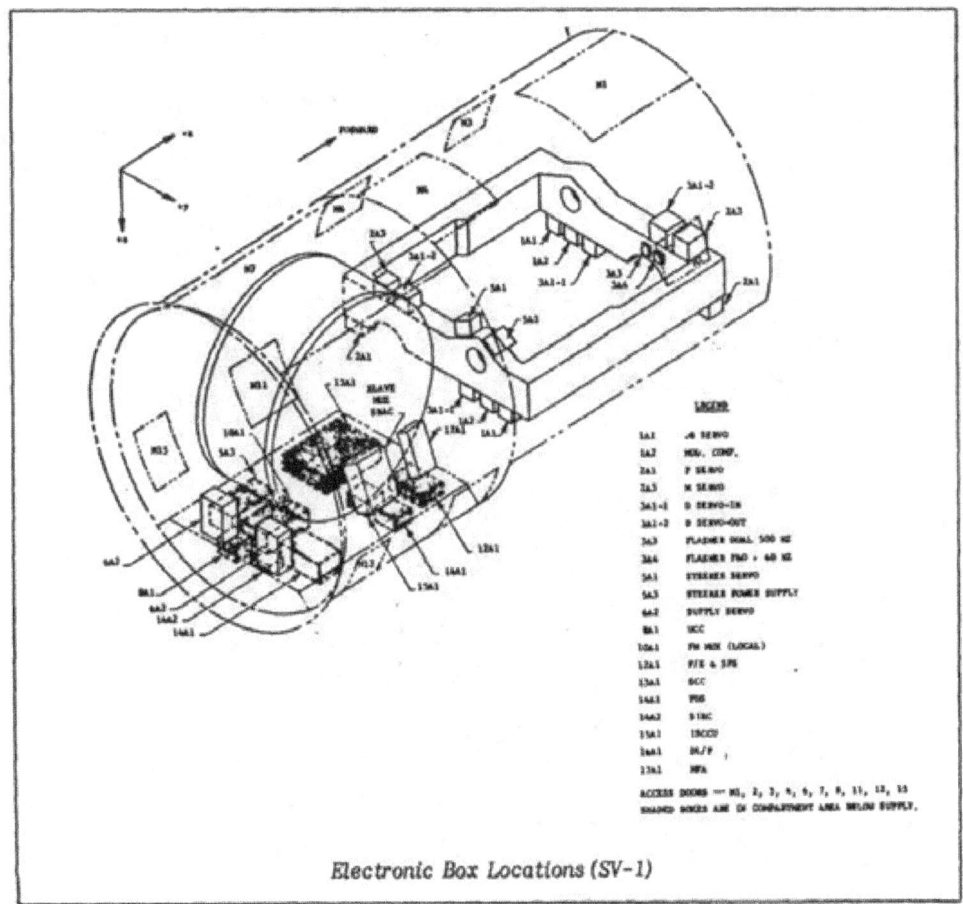

Electronic Box Locations (SV-1)

Assembly compartment.

Although the Perkin-Elmer "Protem" proposal submitted to the CIA in May 1965 did not discuss the details of the electronics system, two reports describing the film transport functions and the on-board diagnostics were written in support of the proposal.[1,2,3] These reports are the basis of the Hexagon electronic designs that followed.[4] An early Perkin-Elmer presentation to the customer included slides showing the location of electronic boxes in the Supply Assembly compartment.[5]

A Perkin-Elmer presentation to the customer on 9 December 1965 was more detailed and included not only a listing of the major electronic circuits, but also discussed the various electrical experiments which had been conducted up to that time.[6]

Electrical design, analysis, and experiments continued up to the day that the Perkin-Elmer Hexagon proposal was submitted on 21 July 1966.[7] The proposal contained a detailed diagram of the control system and listed all the studies, analyses, and experiments which had been conducted since early 1965. In addition, the proposal contained all specifications relating to the electrical design. After the award of contract, the electronics system began to take a definite shape.

The initial electrical design work on the program was started by Robert M. Landsman who early in the effort established some basic ground rules for the people working with him.[8] One of the rules stipulated that every electronic

Cabling Arrangement at Aft End of Midsection

component had to justify its existence in the circuit. "If it ain't there it can't fail." Another rule was, "No potentiometers or relays are to be used in the system." Unfortunately, exceptions to this rule had to be made since it became too prohibitive and less reliable to eliminate these components from some circuits. At least the circuit designers carefully considered the addition of questionable components. Landsman also insisted on AC coupling across all high-level and low-level ground systems to decrease noise.

Initially, the customer requirement was for a system with a fixed scan angle of 120°, each frame being ten feet long. However, soon after the award of contract, it was determined that the system would be more effective if various scan angles could be selected in orbit. According to one engineer, "The short scan saved the system. It would have been a really lousy system if we had stayed with the fixed 120° scan." This was substantiated by the fact that soon after the initial flight of the Hexagon system, the 120° scan was used only for special purposes.

Fortunately, the Perkin-Elmer proposed design lent itself to a change from a fixed 120° scan to various short scans. Perkin-Elmer engineers had designed a 50 millisecond stop at the end of the platen travel. All that was required was to vary the length of this stop. This permitted the mission operators to select a short scan and position it anywhere within the 120° platen travel.

One of the early problems that Perkin-Elmer electrical engineers were confronted with was the method of controlling the platen motion. The platen has to move with the optical bar, recycle, and then lock into the optical bar to repeat the scanning motion. At the same time, the platen had to change its position slightly relative to the optical bar to correct for image motion.

This could be accomplished in two ways. By adding sensors between the platen and the optical bars and doing it electrically, or by using a mechanical arrangement using a three-dimensional cam. Perkin-Elmer recommended the electrical technique, while SETS, the customer consultant, leaned toward the mechanical design and pushed the customer towards their approach. Fortunately for the program, the customer selected the Perkin-Elmer approach.

Another early technical conflict between SETS and Perkin-Elmer was the selection of the sensor design between the platen and the optical bar. The Perkin-Elmer approach recommended in the Hexagon proposal required "E"-core magnetic transducers between the platen and the optical bar. A SETS analysis predicted that the "E"-core approach would not work. Perkin-Elmer then used an alternative approach, capacitive transducers, and were well into the design when SETS came back with another analysis stating that capacitive transducers would not work and recommended a return to the "E"-core approach. At that point, Perkin-Elmer engineers were not about to change their approach again and stayed with the capacitive transducer which operated successfully.

The customer soon gained confidence in Perkin-Elmer's ability to produce reliable electronic designs and work on the electronic system continued without further conflicts. The electronic boxes were reviewed in the same manner as the other functional units. Electrical circuits closely associated with the functional units were approved at the same time as the functional unit PDR's and CDR's.

SYSTEMS ENGINEERING

Although the various Perkin-Elmer engineering groups assigned to particular functional units were cognizant of the overall operation of the camera system and involved in developing interfaces with themselves, other major Sensor Subsystem Associate Contractors, and the Satellite and Re-Entry vehicle contractors, the responsibility for the overall system design of the Hexagon program at Perkin-Elmer was assigned to the Systems Engineering Department.[1,2] This department guided the initial concept design and established performance and the functional specifications for the camera as a system and for the major functional camera assemblies.

Whereas the project engineers of the major functional units conducted design reviews for their specific designs, the system engineers presented technical reviews to the customer on the overall progress of the Sensor Subsystem. In addition, the systems engineers supported the various technical reviews of the individual functional units and published analyses supporting these designs.

Specifically, the Systems Engineering Department was responsible for several areas including: Performance Prediction, System Analysis, System Dynamics, Structural Analysis, Mass Properties, and Thermal Control and Analysis.

During the Sensor Subsystem PDR which was held on 29 February 1968, Donald Patterson, CIA Hexagon Program Manager, opened the review and observed that the Sensor Subsystem PDR constituted the first major milestone in the development of the Sensor Subsystem.[3] He noted that the review served as assurance to both the customer and Perkin-Elmer that the design is viable so that detailing of the design can proceed. He also observed that it provided an opportunity to assess critical problem areas and exchange information that previously (i.e., during functional assembly design reviews) may not have been available.

In his closing remarks, Patterson noted that the design presented at the PDR would be used as the baseline in contractual negotiations with Perkin-Elmer. He observed that there were at least four areas which did not meet the Sensor Subsystem specification as written, and not considered by Perkin-Elmer to be the baseline design. These were: weight, reliability, ability of the Sensor Subsystem to operate within one hour after launch, and the ability to operate within specification on a 0.4 probable day.

About a month later, Kenneth Patrick, Director of the Hexagon Program at Perkin-Elmer received a memorandum from Donald Patterson which included detailed comments on the Sensor Subsystem PDR[4].

It stated, "Despite many favorable comments during the technical consultant's system development program, two areas of concern were nonetheless registered. One was with regard to meeting schedule, and the second was with regard to the complexity of the film transport system, with particular attention to the servo interrelationship, phase-lock loop detailing, and the general complexity of the command and control portion of the sequencer.

It is essential, therefore, that we must focus our attention during the PDR-CDR period to the area which will provide us the best assurance of meeting our schedule with a system of acceptable performance. Major areas on which we must concentrate are system simplicity (output of our design audit), scheduling realism, resolution of high risk design areas, and the performance of critical development tests of critical designs and components." About a year later (March 1969), the Sensor Subsystem CDR was presented to the customer and approved.[5]

Although the Systems Engineering Department participated in all aspects of Sensor Subsystem design, there were several areas which required their special attention. These were pressurization, film tracking, weight control, and reliability. In addition, systems engineers participated on the Design Audit Team which was formed in June 1968 to identify the high risk areas.[6] Of immediate concern were the following: inadequacy of the air bar steerer design, film path pressurization, complexity of sequencer design, uncertainties of servo performance, unresolved problems of the Supply Assembly design, optical component mounting design, and encoder procurement problems.

Pressurization

The decision to pressurize the entire film path required many tests and analyses and much discussion. During the initial stages of the Fulcrum program (November 1965) the requirements specified in a technical memorandum to all project engineers indicated that "During camera operation, it will be necessary to maintain a minimum pressure of 10 μ within the camera compartment to prevent corona effects. During non-operating periods, the pressure may be regarded as approximately equal to ambient pressure at 100 nautical miles or about 0.002 .[7]

Thirteen months later it was discovered that pressurization was more critical to camera operation than initially considered. The first tests of transporting film in a vacuum revealed that vapor and gas released from fresh film wound on a large diameter supply spool caused the outer film layers to float in an unstable manner. It was clear that unless this condition was controlled or prevented, the phenomenon could cause damage to the film and introduce large tracking errors in the film transport system.

A test program was conducted in February 1966 to confirm the existence of control problems due to film outgassing and to develop tracking control measures.[8] The tests were completed on 25 April 1966. A report discussing the results mentions several control devices which were used in the tests.[9] Results varied widely. Success of a control device was determined by the comparative ability of the device to control the film's lateral movement or stability as it moved off the supply spool or onto the take-up spool. The control devices included fences, builder rollers, despooling rollers, self-centering rubber rollers (Lorig-Aligner), and banded spools.

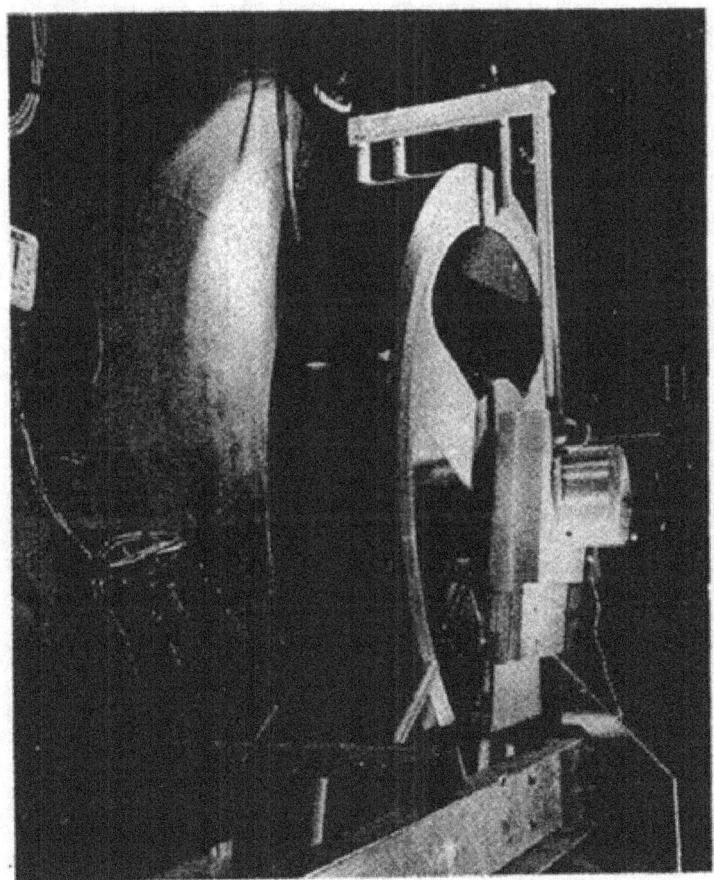

Film Transport in Vacuum Experiments

While stationary fences were satisfactory for the supply spool and for relatively small stacks on the take-up spool, large film stacks rubbing against the fences resulted in unsatisfactory take-up performance.

Four types of builder rollers were used on the take-up spool: banded, flanged, plain, and crowned. All of these successfully prevented large lateral shifts (telescoping) but they performed with various degrees of success in producing a successful stack. The banded roller rated the highest; next came the flanged roller, and the least successful were the plain and the crowned rollers.

Three types of despooling rollers were used on the supply spool: banded, flanged, and plain. All were unstable in controlling the lateral position of the film coming off the supply spool. An important point noted in this test was that the builder roller used in a system requiring rewind will operate as a despooling roller during rewind (supply spool becomes take-up spool). The report recommended that some method be provided to either lift or remove the roller from the film stack during rewind. Tests on the Lorig-Aligner were inconclusive.

The test report made the following recommendations. For a film transport system requiring film rewinding and large diameter spools in an environment that results in film outgassing, stationary fences were to be used both on the supply and the take-up spools to control lateral movement of the film as it leaves the stack. Builder rollers were also recommended with the provision that they contact the film stack only when the spool is used as a take-up and lifted off the film when it is used as a supply spool.

The report also outlined a future test program using full rolls of film, refining the test set-up to obtain quantitative data on the effect of outgassing film on tension control and velocity, and incorporating passive and active steerer mechanisms to effectively control film transport in a vacuum. An analysis of floating film written in February 1966 concluded that about 32 turns of film will float due to outgassing creating a maximum gap of about 2 inches on a 60-inch diameter film supply spool.[10]

The answer to pressurization problems at the time that Perkin-Elmer's Hexagon proposal was submitted (July 1966) were two solutions included in the design of the transport system: one for supply spools and the other for take-up spools.[11]

"Stationary fences provide the best solution for supply spools. A fence is a stationary rod positioned radially to the supply axis and approximately 1/16 inch away from the edge of the film stack. Three fences spaced 120° apart are used on each side of the supply stack. During transport the outer four to five layers of the supply spool will lift, become unstable and shift laterally (telescope) against the fences on one side. The film will ride against or intermittently touch the fences with very small lateral forces. The film lateral position is therefore constrained to ±1/16 inch or the spacing of the fence from the side of the film stack.

The "builder roller" was found to be the best solution for positioning and stacking the outgassing film onto a large diameter take-up spool. This is a roller on a swinging arm which rides on the take-up spool and is positioned such that it rides at or near the tangent point of the film as it approaches the take-up spool. Since a period of film reversal takes place at the end of each photographic cycle, each spool must act as both supply and take-up. Consequently, fences and builder rollers are provided at both ends of the system, with provision for lifting the builder roller whenever a spool is being used as a supply."

About two weeks before the award of contract to Perkin-Elmer, two systems engineers published a memo recommending system pressurization.[12,13] It stated, "Problems associated with film outgassing suggest the need for pressurizing the system. Vapor condensation, film floating force, film temperature and heat loss, and mechanical Q or transmissibility are analyzed as a function of pressure for the nominal operating temperatures ± the three-sigma temperature variation. By choice of the optimum pressure-temperature relationship, the problem of vapor condensation may be eliminated and the other problem areas controlled." A pressure control system was described in the memorandum using temperature-biased pressure relief valves in selected locations.

The first monthly technical report indicated that three approaches to the pressurization problem were being considered and would be studied after the pressure level in the supply assembly was established: (1) a low pressure system (i.e., lower pressure than the supply) (2) a higher pressure system or one that is equal to supply pressure (3) and an adaptive system that is temperature biased and pressure regulated.[14]

By the end of December 1966, the customer consultant, SETS, became involved in the pressurization problem and issued a memorandum describing a program plan to study the problem.[15] Another SETS memorandum voiced concern that the only information available on how film moisture content varies with temperature, pressures, and moisture content of the pressurant gas was the Eastman Kodak Photographic Handbook, "Which is far from adequate."[16] A SETS memorandum dated 4 January 1967 noted that Perkin-Elmer was considering pressurizing the film transport system and presented an initial evaluation of the film and the system associated with a pressurized transport system.[17] It stated that photographic properties of the film were negligibly affected by moisture content and listed a number of specialized problems which required study. Among these was the effect of dry nitrogen from air bars producing local drying of the film and the difficulty of sealing the optical bar.

On 25 January 1967, an informal meeting was held to present both the SETS and Perkin-Elmer concepts of pressurization and to discuss proposed test plans.[18] A Perkin-Elmer systems engineer, in the meantime, published a technical report on the interaction of the film and its environment and again emphasized the need to pressurize the system.[19] The report also contained several recommendations for a pressurized system.

In March 1967, the need for adding a system which would maintain the proper environment for the film transport system was being studied by the systems engineers. At that time, pressurization requirements were undefined and in fact were not included in the budget weight of the Sensor Subsystem.[20] A month later, in a meeting of systems and design engineers at Perkin-Elmer, a decision was made to control the moisture content of the film during the mission.[21] A trade-off study was started to review the various options on film path pressurization. The four options concerning the degree of pressurization included: (1) a completely closed film path with air bars supplied by pump (2) a closed film path with relief valves allowing escape of gas from the air bars supplied by high pressure tanks (3) cassettes, sealed and pressurized, with chutes free-venting and (4) free-venting path with chutes acting as light barriers only.[22]

By April 1967, an experimental work plan for an abbreviated film path to be operated in a vacuum was developed.[23] It included tests to evaluate film path pressure and pressure control devices, including chutes, relief valves, cassette seals, and pressurization apparatus. In addition, a film flatness test was planned along with an investigation of film flutter effects introduced by the twister operation in a pressurized system. This was followed by a decision to provide pressurization of the supply assembly only using a resilient gate which closed around the film strip during shut-down. It listed various reasons why a completely pressurized film path was not only difficult to implement but created other problems. A memorandum supporting this decision included a complete bibliography of all program documents discussing the pressurization problem.[24]

In a technical meeting between SETS, the customer, and Perkin-Elmer on 31 May 1967, the customer requested a description of the basic rationale behind the decision to pressurize only the Supply Assembly. The customer rejected the above memo as an inadequate basis for a decision on system pressurization and asked for a definition of the testing or analytical data that would permit a pressurization decision prior to CDR. The meeting resulted in several action items imposed on both SETS and Perkin-Elmer.[25]

Pressurization studies continued, but in the meantime the design of the Sensor Subsystem continued based on an unpressurized film path. In July 1967, the monthly technical report included concern over film curl. "The danger of film curl causing film contact with the chutes in the long, narrow unpressurized chutes will require further investigation."[26]

Ten months later, spooling tests on the abbreviated film path equipment revealed that film spooling problems in the take-up occurred during the rewind operation.[27]

SETS published a memorandum (16 May 1968) indicating concern that some Perkin-Elmer engineers did not fully agree that pressurization of the entire film path would reduce the loss of moisture from the film during orbital operations. It described the manner in which pressurization would reduce moisture loss and the problems that this would alleviate. These problems included corona, take-up ballooning during rewind, film curl, loss of film stack integrity in the take-up water expansion from the supply into the film chutes, low film temperatures in the film path due to outgassing, and film contraction in the chutes during dormant periods.[28]

This was followed by another SETS memorandum (11 June 1968) that reemphasized the need for a pressurized film path design as a backup capability. However, the memo stated that conversion of the baseline film path to a pressurized design was not recommended until further definition of the pressurized system was achieved. "The schedule slip involved in the conversion will be minimized if the pressurized design is aggressively pursued as a backup now."[29]

By June 1968, Perkin-Elmer systems and design engineers were convinced that a pressurized system was necessary, as indicated in a monthly technical report to the customer, "It thus appears that if the problem of ballooning in the take-up is to be avoided by pressurizing the film path, then leakage of gas from the film path during periods of nonoperation must be reduced."[30]

Finally in September 1968, a decision was made to pressurize the entire film path. This was reported in the monthly technical report which stated, "A decision was made to seal the entire film path so as to maintain film moisture in equilibrium. Maintaining a partial pressure of water vapor in the system not only conserves film moisture but also inhibits ballooning of the film during rewind. Reduction of curl-induced focus error provides still another benefit."[31]

Since by that time, almost all of the major assembly CDR's were presented and approved and detail design was substantially completed, a major effort was required to redesign those units affected by the decision to pressurize the film path. These included all of the Film Path Assemblies, the Looper Assembly cover, the Optical Bar seal, the Platen and Film Drive Assembly covers and seals, and the Take-Up Assembly.

The decision to proceed in this manner was based on the necessity of maintaining the Hexagon program schedule. If back-tracking was necessary due to a lack of test data which prevented an early decision, it was a price that had to be paid for developing a camera to meet the stringent requirements of the Hexagon program.

Film Tracking

Perkin-Elmer's philosophy of transporting film in the Sensor Subsystem was initially stated in the Hexagon proposal.[32] "As has been made clear in the discussion of design optimization, a certain degree of complexity, and a certain burden of effectiveness has been placed upon the film transport system in order to acheive a maximum of simplicity and reliability in other system areas. This has been done in the light of a background of experience in highly precise film transport systems for panoramic cameras which provides assurance that the problems which exist here are problems which have been faced before, and for which effective solutions have been found.

Experimental programs which have been carried out to explore areas where the pertinent parameters for this system exceed the range of values applicable in prior developments (e.g., film velocity, mass of film supply, operation in vacuum) have confirmed that the methods which have been previously developed are adequate for the present situation."

Prior to the Hexagon program, reconnaissance camera systems had a relatively slow film transport speed and no reversals of film direction. The Hexagon design approach proposed by Perkin-Elmer and accepted by the customer required a high speed film transport system (204 inches per second maximum in the fine film transport and 68 inches per second in the coarse film transport) with film reversals both at the take-up and supply spools.

In addition, the length of film between the supply spool and the take-up spool was a maximum of 100 feet. Combined with the fact that the film supporting elements (rollers, airbars, film spools) were independently mounted either on the vehicle structure or on the camera frame resulting in possible assembly misalignments and misalignments due to launch and orbital operations (i.e., thermal causes), the Hexagon film transport system had the potential of experiencing severe tracking problems due to testing and mission operations. The answer to these potential problems was briefly stated in the proposal. "The components which affect film path alignment are mounted in associated units and interfaced with vehicle structure in a manner that will preserve initial alignment."[33]

The proposal recognized that film handling was a development risk, "The present system represents a step beyond the current state-of-the-art in film handling in that a very large supply of film is provided, the film is unusually thin, it must be handled at relatively high velocities, and it must operate in a zero gravity and zero pressure environment."[34]

The test bed for the Sensor Subsystem film transport system was the Film Path Simulator (FPS). It was composed of breadboard-type supply and take-up spools, a looper, tension sensors, drive and metering capstans, and a film platen. The major objective of the FPS was to simulate the dynamic characteristics of the film transport system.

Some of the initial problems of film tracking were identified in the first monthly report (December 1966). "The FPS has been completely aligned. UTB film has been handled successfully, but all perturbations (localized flutter in the web due to dynamic tension gradients) cannot be removed without major rework of the film drive and platen assemblies."

Initially, the FPS was operated without the looper shuttle and film drive assemblies in order to minimize the number of elements that could create film disturbances. It was found that when properly aligned, the skew bars could operate as self-aligning bars in both directions.

With the platen and film drive assemblies added to the film path, excessive film perturbations and lateral film travel appeared at the slit area. Since the film path without the platen and film drive assembly was relatively free of film disturbances, it was concluded that the perturbations were caused by errors within the film drive and platen assemblies.

Upon further examination, the capstan assembly was found to be responsible. Particularly the torque motor shaft of the capstan assembly was not properly assembled to the capstan. This assembly was replaced in the film path. A marked improvement resulted. No visible disturbance appeared. Trouble-free operation was accomplished with UTB film.

The twister and platen assemblies were also checked on the surface plate. Elements of these assemblies were cleaned and realigned. The twister and platen assemblies were then replaced in the FPS. In spite of the re-alignment of the assemblies, film perturbations still appeared. Thus, it was evident that the residual errors in the twister and platen assemblies were limiting performance.[35]

A preliminary film tracking analysis was completed in May 1967 and later expanded to a technical report and published in November 1967.[36] The report developed a mathematical model "which could be used to predict the response of the film medium in passing between two adjacent misaligned rollers."

In December 1967, an analysis was completed which "verifies the compatibility of the film tracking portion of the functional specification on interconnecting film path assemblies with current design concepts and film tracking theory.[37] An additional analysis was completed in February 1969 and presented a discussion of the elements of the film path and their contribution to film tracking errors. The report also included references to previous film tracking analysis and system requirements.[38]

Throughout the program, test engineers and technicians working with various simulators and actual hardware were continually faced with tracking problems. These were investigated and adjustments were made to correct tracking for that particular situation.

On 14 November 1969, the Engineering Model was installed into Chamber A. Full system operation was achieved in the chamber.[39] However, intermittent film tracking necessitated removal of the Engineering Model which was set up outside of the chamber to determine the cause of the problem.

Film tracking tests were conducted at constant velocity to verify tracking in the chamber film path and to confirm system operation. It was necessary to make minor alignment adjustments to the chamber film path under dynamic conditions. Good tracking was achieved at constant velocity. A film jam subsequently occurred in the Camera A fine film path of the Engineering Model. This failure was caused by the instability of the Chamber A film path, mounted to the chamber floor, which passed the film through a chamber access point to a takeup on the outside.

The Engineering Model was reinstalled in Chamber A. Good tracking was observed until the scan angle was changed to 120 degrees, whereupon tracking deteriorated badly. Testing was suspended and an inspection revealed a film jam in the camera.

In that same period of time, the Development model reached a point at which film transport system tests were started. On 4 November 1969, the coarse film path B was threaded and operated successfully. The fine film path on Camera B was then threaded and spliced. The system was operated and a problem in film tracking was noted. The simulator supply was offset an amount that allowed the proper adjustment in the film path at the crossover assembly.[40]

Meanwhile, the Flight Models began approaching the critical point at which their film transport systems would soon be in operation.[41] During initial film tracking tests in the Flight Model 1 (S/N 002) midsection, severe film wander was noted on both A and B film paths. Diagnostic and visual observation determined that the film was wandering at the supply film exit vestibule at a once-per-revolution of the supply. The design of the aft articulator had been changed since previous runs, due to qualification test failure of the articulator. Therefore, one of the new articulators (A-side) was removed and the old design reinstalled. There was no improvement initially, but after a total of 3000-4000 feet of film had been transported, the film wander suddenly disappeared and did not recur. The new design articulator was reinstalled and tracking remained good. Similarly, the B-side tracking improved after 3000-4000 feet. Thus, it was concluded that poor tracking was associated with a poor outer section on the film stacks. This was later to be attributed to a change of film from Type SO 380 to Type 1414 which had a taper and caused film spillage after a buildup of a high number of turns.

On 10 July 1970, the decision was made to replace S/N 002 (Flight Model 1) with S/N 003. This decision was based on a preliminary failure analysis of the 47° test run (failure of a component in an electronics box) in which it was determined that extensive disassembly and rebuilding of S/N 002 was required.[42]

S/N 003 (now Flight Model 1) was prepared for vacuum testing. On 18 August 1970, during maximum rewind, a jam developed in the B Camera. Attempts to clear the jam through system operation were unsuccessful, so the chamber was vented.[43] By 19 August, the B-side jam was cleared in place and system anomalies were investigated. The cause of the jam on the B-side was a broken wire in the take-up steerer B. Repairs were made and on 28 August the 47° vacuum testing commenced. The Supply A steerer soon showed a saturation condition, followed immediately by a film jam in the A fine film path. All attempts to clear the jam through system operation were unsuccessful, so preparations were made to vent the chamber. System investigation revealed no film stack anomalies. An intensive investigation of the cause of the jam was initiated but revealed no anomalies in the steerer electronics or the film stack.[44]

By 20 September, several assemblies were replaced on Flight Model 1 and then it was installed in the chamber and rerun. The 70°F vacuum tests were aborted due to a film jam on both sides. Subsequent investigation revealed that the command and control box was not the cause of the failure.

On 16-23 September 1970, a Film Tracking/Alignment/Servo Committee review was held at Perkin-Elmer at the direction of the Director, Photo Reconnaissance Systems, Office of Special Projects. The committee was formed for the purpose of reviewing the history of metering capstan, platen, and film tracking problems on the Hexagon camera system and to identify the causes of the continuing problems. There were twelve members on the committee from the Special Projects Office, Lockheed, Aerospace, and SETS. Several action items and technical directives were recommended by the committee.[45]

A memorandum from Patterson to Maguire (Perkin-Elmer Program Director at that time) stated that, "Even though a root cause of the repeated film tracking problems was not identified, several weaknesses in the area of analysis, test, and procedures were evident."[46] As a result of the review, several analyses on film tracking were written and tests were conducted.

By that point in time, the engineers and test operators were becoming more alert to any problems related to poor film tracking. On 3 March 1971, a baseline test on Flight Model 1 experienced a stacking problem on the A side. It was the first indication that something other than and equipment failure was the cause. The film stack wedge was measured and found to be 0.021-inch high outboard. Film samples were measured and were also thicker outboard.[47]

In April 1971, during post-Chamber A-1 inspection of Flight Model 1, a film foldover was noted on RV-1 take-up A. The foldover occurred about halfway through an inadvertent 47-minute run and corrected itself without causing an emergency shutdown. Film wedging was determined to be the cause of the problem.[48]

Tracking and stacking problems were also occurring on Flight Model 2 (S/N 002) and Flight Model 3 (S/N 004). But by May 1971, just a month before the launch of the first Hexagon Camera System, tracking seemed to be under control on all operating Flight Models.

Perkin-Elmer engineers had at last developed an assembly technique that seemed to contribute to the good tracking that was now being experienced. Tracking tests on the cameras followed a plan which progressively built up the film path; tracking was checked at each stage.[49]

During May 1971, an investigation was also undertaken to determine the effects the crossover airbar adjustments had on film tracking. A film taper test was also planned on the Engineering Model to start in the early part of June.[50]

Flight Model 1 was launched successfully on 15 June 1971. There were no tracking problems during the mission.[51]

In July 1971, during testing on Flight Model 3, marginal stacking was noted on RV3 for the first 8000-10000 feet. A concave condition of 0.006-inch and a film wedge of up to 0.033-inch were measured. From 10,000 feet on, the concavity gradually reduced to zero and the wedge came to less than 0.005-inch. Tracking and stacking were exceptionally good. Subsequently, a manufacturer's splice was found at the point that the off-track began. No further problems were encountered.[52]

Tests on Flight Model 4, however, exhibited severe film wander on Camera B.

The problem occurred during test Sequence 140. Later at the start of Sequence 141, the Sensor Subsystem Test Console issued an emergency shutdown command due to the supply B steerer being out of limits. In each case, the rewind speed preceding the above sequences had been 55 inches per second. The sequence was restarted and ran with no additional problems. All subsequent testing was successfully completed.[52]

No serious tracking problems were reported on any Flight model ground tests until October 1971, when Flight Model 6 experienced tracking problems due to a faulty edge sensor. There was also a film tracking problem at the same time during start-up. Subsequent shimming of the frame articulator finally restored proper tracking.[53]

Flight Model 2 was launched on 20 January 1972. Although the mission was completed without film tracking problems, there was a film separation on the B side of RV-3 caused by film sticking due to contamination introduced during the film manufacturing process.

Six months later (7 July 1972) Flight Model 3 was launched. Two serious tracking problems occurred. The first was the occurrence of a film foldover in the Aft Camera film at the initiation of rewind. Since the Aft Camera fold was being generated in RV3 take-up, Aft Camera operations were resumed on RV4 take-up. The second problem was another film foldover which occurred on the Forward Camera during rewind operation. Since this fold occurred on RV4 take-up, stereo operations were not ceased until the take-up was nearly full.[54]

This failure prompted the formation of a Tracking Task Force to investigate the causes of Flight Model 3 tracking problems. To provide a background for task force activities, a report was prepared which summarized film path tracking investigations through the history of the Hexagon program. The task force was composed of eight members; two from the Special Projects Office, three from SETS, one from Eastman Kodak, and two from Perkin-Elmer.

The nature of the failures on Flight 3 was the subject of an extensive study by the 1203 PFA Team, however, they were unable to identify the exact mechanism of the failure.[55] The nature of the 1203 failures was that on two occasions a disturbance occurred in the coarse film path which resulted in an automatic Emergency Shutdown of the camera system. This was followed by evidence that a film foldover had occurred and was being spooled onto the take-up. In the first instance, operations were discontinued on the Aft Camera, while RV3 take-up was in use, because of concern that continued operation would result in a film path jam which would prevent the transfer into the fourth RV.

The second instance, which occurred on the Forward Camera while operating into RV4 take-up, resulted in the system being operated in an inefficient but operational mode. By wrapping folded film onto the take-up, the maximum radius was reached earlier than would have been the case without a fold, reducing the amount of film returned. The Tracking Task Force investigations continued for three months (September through December 1972). After conducting numerous tests and analyses, the task force made recommendations placing constraints on the operation of Flight Models 4 and 5 and recommending a retrofit on Flight Model 6. Many other recommendations were made including builder roller investigations, collecting a library of film samples known to have caused film spills, and instituting a routine measurement program by the film manufacturer for determining the profile, rather than the taper, of each film segment used in a flight film roll.[56]

Flight Model 4 was launched on 10 October 1972 and Flight Model 5 was launched on 9 March 1973. Both camera systems were successfully operated.

Two days after mission 1205 (Flight Model 5) was completed (6 June 1973) Patterson wrote a memorandum to Michael Weeks (OTD General Manager) expressing a concern about the tracking problems on the Hexagon Sensor Subsystem.[57]

"We have always had tracking problems in the system. Numerous things have been proposed as the casual elements and some have been tracked down and corrected. However, the tracking problems are still with us and seem to be getting no better. Serious problems in tracking on SV-3 resulted in tracking constraints on SV-4 and SV-5 which caused a loss of about 15 percent of our collection capability. Translated into dollars this represents a loss of about ▓▓▓▓▓▓.

From the problems we have had it appears to me that we may well have a generic design problem with the film path which permits the tracking to change without our understanding why. Certainly, the alignment is still a "black art" when only one or two people are capable of aligning a system and then only by trial and error. It seems to me that the tracking should be amenable to a systematic alignment procedure that any competent technician could follow. I think we should look at alignment adjustments for various elements of the path which will permit ready alignment with standard procedures rather than the twisting and turning of elements as is now done."

Weeks replied to Patterson's memorandum and indicated that Perkin-Elmer was also concerned about system tracking and was developing improvements and incorporating them into the system as quickly as possible.[58] "First let me address the tracking problems on SV-3 and the resulting rewind constraints on subsequent models. We are convinced that the problems on both sides of SV-3 were a result of film having different stacking characteristics from the film we had used before and were not due to any camera problem. We developed a new builder roller design which should be more tolerant of film variations such as crown and taper. Although we have tried to implement this improvement on a crash basis, there seems to be some reluctance on the part of the government to carry out this program."

Weeks noted that Patterson's comments regarding the "black art" of tracking and alignment were well-founded and recognized early in the production process. Changes were made to permit ease of adjustment. By adding fences on the supply assembly, it was possible to rewind the film up to 80 inches per second on Flight Models in Block II.

Weeks concluded his reply by stating, "We feel that the above changes have solved most of the tracking problems exhibited by the early models." Apparently the modifications made on subsequent flight models solved the tracking problems since Flight Models 6 through 10 had no tracking problems that were not easily corrected.

This success, however, was short-lived. Ground tests on Flight Model 11 during the end of 1975, just prior to launch, was troubled by tracking problems in the fine film path. Rebuilding and adjustments of critical assemblies did not help.[59]

Tests were conducted on the Engineering Model in an attempt to determine the cause of mistracking on Flight Model 11. In parallel with the Engineering Model tracking tests, the physical properties of the mistracking film were being closely examined. A microscopic examination of the mistracking film revealed an extruded projection at the edge of the film which was subsequently identified as being caused by the slitting knives used during film manufacture. The film in the same stack without the extruded edge tracked correctly.

In 1976, when tests were being conducted to qualify SO 208 film for flight use, extremely bad tracking was encountered. It was verified that matte particle size (pelloids), together with their distribution, had a direct influence on film tracking. Fine particles led to a lower coefficient of friction with the subsequent lowering of lateral tracking stability.[60]

One of the early difficulties in identifying and correcting film tracking problems in the initial stages of the Hexagon program was that there was no single cause. Poor tracking could be caused by film supporting elements and their structures, improper alignment procedures, and film error contributions such as wedge, splices, damaged film edges, and material composition. The more subtle tracking causes were masked by the more significant. As each problem was eliminated, others remained to cause mistracking.

Perkin-Elmer engineers and technicians now have the experience and knowledge to align film transports using standard procedures, and make the necessary modifications or adjustments to eliminate tracking problems. In addition, film physical properties are carefully examined to assure that the material meets transport requirements.

It is unfortunate that the overwhelming complexity of the film properties and the film tracking problems prevented resolution in the early stages of the program. Perhaps the readers can apply the lessons learned on the Hexagon program to problems that they may face on future programs.

An exact knowledge of not only the photographic characteristics of the film but also its physical properties was of importance in other areas of the Sensor Subsystem in addition to film tracking. Although a meeting with Eastman Kodak in the early phase of the Hexagon program provided Perkin-Elmer with information on most of the properties of the film, some information was not readily available.[61]

As on other camera programs, design of the film was developed to suit the photographic requirements of the camera mission and the manufacturing processes of the film supplier. The resultant physical properties of the film were initially of secondary importance to the film manufacturer.

In order for Perkin-Elmer engineers to make a decision on whether to pressurize the film path, it was important to know how film moisture content varied with temperature, pressure, and moisture content of the pressurant gas. It would not be until September 1968, after sufficient tests and analyses were completed, that a decision to pressurize the film path was made.

One additional area of concern that surfaced about the end of 1968 was the thermal problem in the forward section of the satellite vehicle. An October technical monthly report in that year states, "A subject that needs more emphasis at this time is the integration of all thermal requirements in the forward section. Thermal design criteria being used by SBAC, MWC, and SSC should be consistent.

Also early agreement is necessary on the method of maintaining the forward section sufficiently warm relative to the midsection so that water vapor from the midsection does not condense in the forward section film path."[62]

In April 1969, Perkin-Elmer established the requirements of the Active Thermal Control System. In addition to heaters which would be located in each RV, thermal insulation was required on the exterior of the articulator housings and chutes.[63]

Finally, in September 1969, interface requirements between the Sensor Subsystem, Satellite Vehicle, and Re-entry vehicle contractors were agreed on and a Perkin-Elmer project team was established to expedite the development and fabrication of Sensor Subsystem equipment for the Active Thermal Control System (ATCS).[64]

There were no problems in designing and fabricating the electrical and mechanical portions of the ATCS. Perkin-Elmer, however, was never faced with having to develop new sewing techniques for making and tailoring insulation blankets

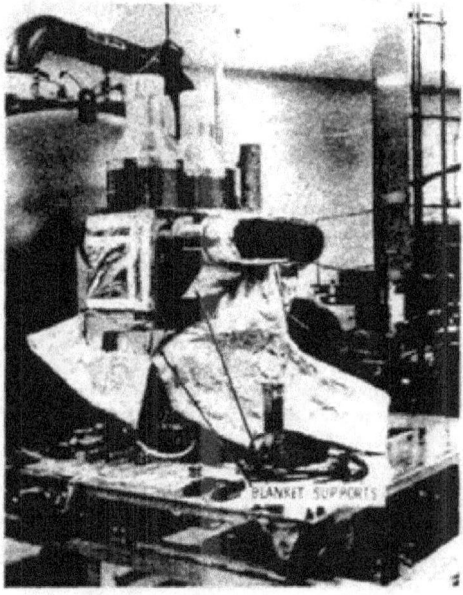

RV4 Thermal Insulation Blanket (Partly Removed)

which were made up from 1/4 mil thick aluminized mylar. The system was tested and proved to be successful in operation.[65]

Weight Control

It appears that a definite pattern in the sequence of events is followed in the weight history of most payloads. Proposal engineers usually estimate the weight of a payload based on a concept and preliminary drawings. They are also, of course, influenced by the weight requirement specified in the RFP.

After the contract is awarded, usually months after the proposal was submitted, an initial weight estimate is made. But by this time the engineers may have spent some time producing more complete drawings and have a better understanding of the components that will be required. In the meantime, the customer may have changed the payload requirements slightly and requested design changes.

It is not until a weight engineer is assigned to the program that a more realistic weight estimate is made. His past experience and knowledge of the items that are usually omitted, such as small hardware, cables, and redundant components and assemblies, results in a weight estimate much greater than everyone expected and what is acceptable by the customer.

This leads to the next step in the pattern, a weight reduction program that concentrates on the use of lighter materials and optimization of the material in the present design. This usually results in an immediate drop in the estimated weight of the payload. Soon after, however, there is a slow but continual rise in the weight estimate in spite of any efforts to keep it down. At this point, a weight review board is usually established.

What may be happening to cause this increase are customer-requested changes, changes to provide a better design, the need for mechanical or electrical redundancies not anticipated in the proposed reliability estimate, redesigns due to part or assembly failures.

Not until all parts are detailed, and estimated and calculated weights are replaced by actual weights, does the payload weight approach what may be the final number.

What is of concern to the payload program manager of today is that this same pattern persists on current programs. "Until the proposal managers assign a weight engineer to participate in the initial concept and he is assigned to the program when the contract is awarded, weight increases will continue to create program difficulties," stated an engineer who participated on the Hexagon program. "A weight contingency must be set aside at the beginning of every program and meted out in an organized manner. Sophisticated aircraft and spacecraft companies and government agencies follow this procedure today."[66]

The Hexagon program is an example of a payload that followed this classic pattern. The weight of the Sensor Subsystem proposed in 21 July 1966 was 2997 pounds. Immediately after the award of contract, an initial weight estimate was made. However, three months had gone by during which time the engineers developed more complete designs and had a better understanding of the components that would be required.

By February 1967, the estimated weight increased to 3847 pounds and finally reached 4066 pounds in March 1967. This was the point at which a weight reduction program was started. By July 1967, the weight dropped to 3821 pounds.

A significant change was made at that time. The customer now required four re-entry vehicles instead of two. The weight estimate shot up to 4308 pounds and by January 1968, when the weight reached 4513 pounds, a new weight control plan was initiated.

Additional customer-required changes, mandatory design changes due to part and assembly failures, and the need for additional redundancy, all resulted in an upward trend of the weight estimate. In May 1968, a weight review board was established to control this situation.

The decision to pressurize the film path resulted in an additional increase that forced the weight up to 4904 pounds. The

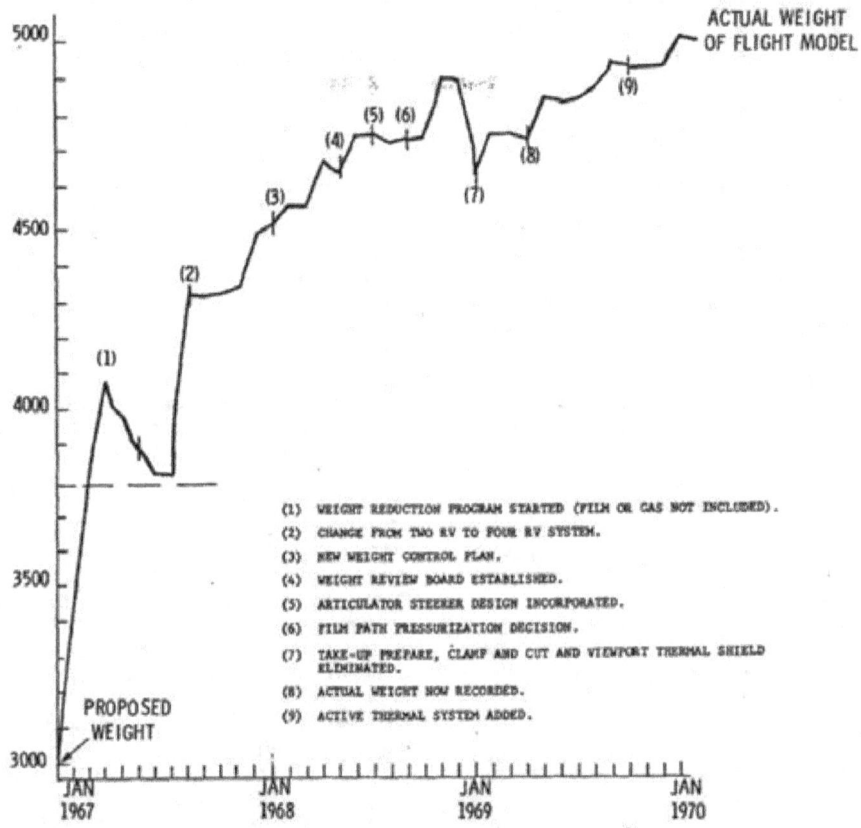

Weight History of the Hexagon Sensor Subsystem

weight number now included a greater percentage of actual part weights.

Soon after, it was decided to include an Active Thermal Control System in the forward section of the Sensor Subsystem film path. The weight increased to 4998 pounds in November 1969. About 92 percent of this weight number consisted of actual part weights. The final weight of Flight Model 1 was 4968 pounds (not including film or pneumatic gas supply).

Film Path Test and Analysis

One of the most useful and effective pieces of test equipment developed for the Hexagon Camera was the Abbreviated Film Path (AFP). The equipment was primarily used in the beginning by systems engineers to confirm their analyses of the film transport system both in air and in a vacuum.

Plans for building the AFP test equipment were initially discussed in January 1967.[67] The objective of the experimental program was to evaluate the physical effects of variations of environment on the interfacing film and film path components. Initially the size of the vacuum chamber was to be 6 feet x 8 feet and the test equipment was to be fabricated and assembled in 18 weeks.

In February 1967, a review of the instrumentation required to obtain thermal data from the AFP was completed. The instruments selected would provide information on film cooling, film outgassing, roller cooling, and film heat transfer characteristics.[68]

A preliminary experimental work plan was written for the AFP by the beginning of April 1967.[69] The tests were to be run in three phases. Phase A was to be the evaluation of film path pressure and pressure control devices, film chutes, relief valves, cassette seals, and pressurization apparatus. Phase B was to include film flatness tests on the platen design. Phase C was to provide test data on film flutter effects introduced by the twister assembly during operation in air and in vacuum.

However, it was not until July 1967 that drawings for the AFP were completed, parts ordered, and fabrication started.[70] The AFP was completed in January 1968, just about the same time that the 10 foot x 12 foot vacuum chamber (now E-Chamber) was delivered to Perkin-Elmer.[71] The Hexagon program was still located at the 77 Danbury Road facility in Wilton, Connecticut during that time.

The initial pumpdown tests began in February 1968 and debugging activity on the equipment continued until April 1968. But by June 1968, the AFP was in operation and tests were being run on film ballooning on the spools. The test data proved that pressurization of the film path was required. This was the first major milestone on the AFP.[72]

By October 1968, Phase A of the test plan was completed and Phase B film flatness tests were started. In addition, to tests conducted to determine the cause of the film corrugations at the twister airbars, some rewind tests were continued.[73]

In November 1968, the AFP was moved to Danbury. This was the last piece

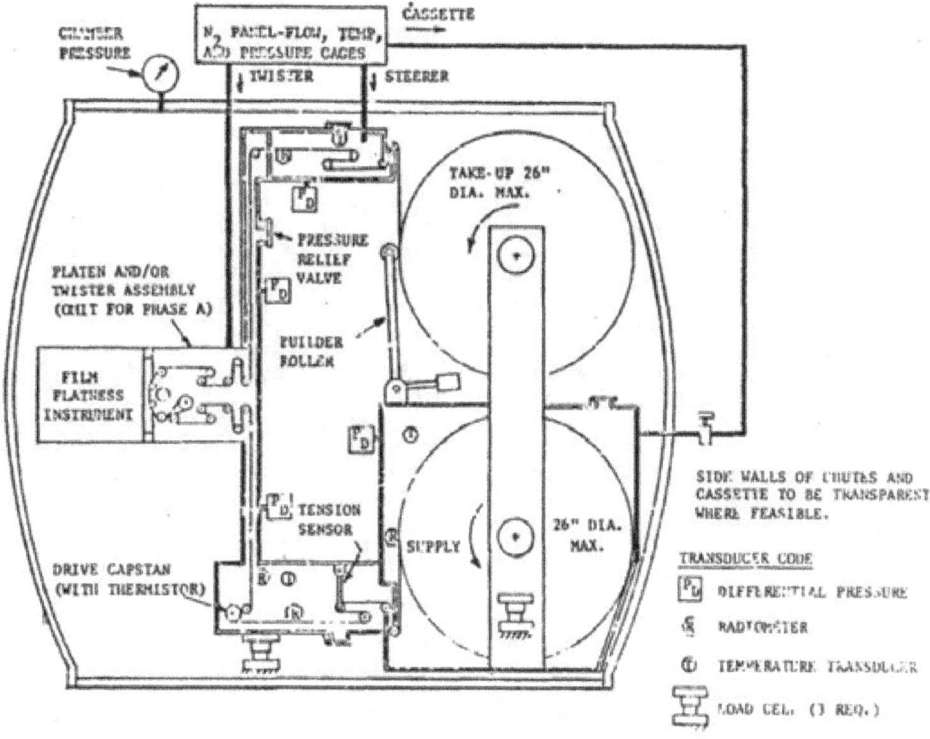

Abbreviated Film Path Schematic

Installation of the AFP into the Vacuum Chamber

of equipment moved from Wilton to the new building.[74] After the AFP was set up, film flatness tests continued, in addition to some corona and film sticking tests.[75]

Phase C tests included the investigation of film flutter effects in the twister assembly. After these Phase C tests were completed, film flatness tests were again run and continued past the launching of the first Hexagon system.

When the AFP was first assembled it contained a takeup and supply, tension sensors, drive capstans, platen, film drive, and a crossover airbar at the supply. Articulators were added at a later date.

In 1974, the AFP was modified to permit film recycling. Prior to that change, the AFP was a constant velocity machine (from zero to 240 inches per second, forward and reverse). The "B" side Engineering Model Looper Assembly was added to the AFP, in addition to the necessary electronics to operate the added equipment.

An engineer who started working on the AFP in 1967, recalls one of many incidents in which the AFP played a significant part.[76] "A couple of years ago (1979-1980) just prior to the launch of a Hexagon system, test data at the West Coast facility indicated a 15-20 micron out-of-focus condition on one camera. AFP tests, however, showed that the test film had a 20 micron dip in the center. Plans to change the platen and add a tilt into it were abandoned since this indicated that the Hexagon Camera was not at fault. Since then, a decision on the final camera adjustments just prior to launch is not made until AFP film flatness test data is examined."

The AFP is currently used to test any new film developed for the Hexagon Camera to check film stacking characteristics, film sticking, film splices, and film flatness profiles.

This same engineer has a great deal of respect for Eastman Kodak personnel. "You have to give those guys credit. We put their film through quite a bit -- twisting it, zinging it back and forth, accelerating it up to eleven miles per hour, stopping it, instantaneously, and reversing it eleven miles per hour in the opposite direction. The stuff is only 1½ mils thick and it's twisted, bent and pressed. In addition, the film coefficient of friction must be correct since the Hexagon Camera doesn't even have sprockets or flanges -- it's amazing!"

SYSTEM RELIABILITY

In the early 1960's, prior to Perkin-Elmer's involvement in the Hexagon program, the reliability activity at Perkin-Elmer was primarily reliability assurance. This effort was basically a review of program test procedures and plans to assure that the required tests were conducted properly. In addition, reliability assurance engineers analyzed failures and malfunctions.

Reliability on the Hexagon program was expanded to cover all aspects of reliability including reliability engineering, which required involvement in the initial conceptual design and design reviews.

Perkin-Elmer's Hexagon proposal expressed the reliability philosophy followed on the Sensor Subsystem. "The mechanisms for computing the reliability of a system composed of electronic and mechanical parts are well established. These are based upon the mathematical combination of failure probabilities of individual components. The guide to the application of these mathematics being the established failure rates of the components involved and their arrangement within the system. This exercise yields a predicted probability of failure for the overall subsystem."[1]

This philosophy was based on the assumption that there were no undiscovered design or workmanship errors in the subsystem. To eliminate these problems, a comprehensive testing program was established to uncover design errors, repetitively occurring workmanship errors, and some isolated nonrepetitive workmanship errors. Errors that remained undiscovered throughout testing and capable of later failures were considered major components of "random failures."

The foundation of Perkin-Elmer's Hexagon part program was borrowed from the Minuteman Program. However, as discovered in later years, Hexagon parts operated beyond expectations since they were of better quality and more reliable.

The Reliability Department peaked at about 65 people. This did not include the personnel working for subcontractors. In addition to setting up a reliability training course for all Perkin-Elmer engineers on the Hexagon program, the reliability group also established reliability programs at subcontractor facilities.[2]

The customer and Perkin-Elmer management were committed to producing a reliable system — cost was secondary. As expressed by Stanley C. Karachuk who managed the reliability effort on the Hexagon program for many years, "We did what had to be done to assure the highest reliability possible."[3]

To accelerate the reliability engineer's learning curve, malfunction reporting was started even before the contract was awarded. "We probably had the most thorough failure analysis activity that was being done at that time," stated Karachuk during an interview. "We were also leaders in promoting reliability techniques and an approach which limited the variety of parts that could be used on the system."

The efforts of all the functions in a reliability department are directed to the production of a reliable system. These efforts are reflected in a reliability number (from zero to one) which is used as a measuring tool. "However, we didn't make it a numbers game," said Karachuk. "We emphasized the design support activities and influenced the design at the conceptual and preliminary design stage. The reliability number was used primarily to compare the effectiveness of competing designs."

The Perkin-Elmer proposal estimate of system reliability was determined to be 0.8819, based on the number and type of parts that comprised the design at that time.[4] Between the time the proposal was submitted to the government (July 1966) and February 1967, the reliability number decreased to a low of 0.6671. This dip reflected a buildup of complexity as the designs progressed. Circuits that were initially estimated at 10 parts increased to 15 parts in the preliminary designs. It was obvious that the reliability estimate in the proposal was based on a very simplistic understanding of system complexity.

At this point, the reliability engineers became very concerned because the low reliability number represented a very high risk that could discourage the

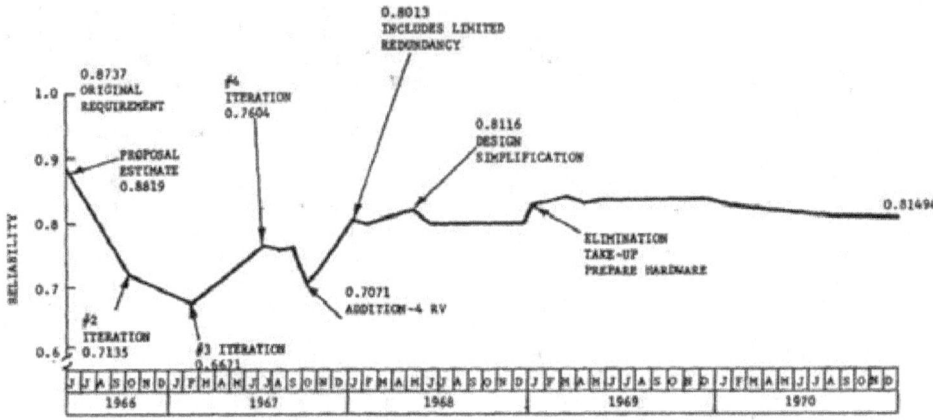

System Reliability Profile History

customer from continuing the program. Additional effort was made to simplify the designs and decrease the number and variety of parts that engineers could use in their designs.

By the fourth iteration of Sensor Subsystem reliability, the estimate increased to 0.7604. This was due to the "get well" program initiated by the Reliability Department. Redundancies were also designed into the system. A second drop in the reliability number (0.7071) occurred in November 1967 because of a change from two reentry vehicles to four, the addition of a variable scan, and the inclusion in the reliability estimate of all diagnostic instrumentation.

Fifteen months after the start of the program, the reliability number was up to 0.8013.[5] This was accomplished by limited redundancy. Beginning in August 1968, the reliability estimates which had been based on in-orbit operation only, now included the entire mission from launch to completion.[6] As the design took final shape in 1969 and the CDR's of each functional unit were approved, the reliability estimate tended to be almost constant.

Although reliability estimates were no longer reported in the monthly reports after the first mission, and the effort became primarily a reliability assurance activity, the overall reliability of the system continued to be monitored in later missions. The failure rates experienced in the later missions were primarily operational failures. The reliability of the Hexagon system has actually increased because the initial design safety margins were much greater than anticipated and Perkin-Elmer screening techniques proved to be much more effective.

On 21 July 1978, Robert H. Sorensen, President of Perkin-Elmer at that time, received a letter from Major General John E. Kulpa, Jr., Director of the Office of Special Projects, Air Force.[7] "The Hexagon Mission 1214 (14th flight) Panoramic Camera is the latest in a series of electronics problems experienced on the Hexagon program. I am concerned that this failure, along with other mission anomalies, may indicate a significant deficiency or even deterioration of the reliability of the Hexagon program. Because of this concern, I am directing an extensive and independent audit of the Hexagon payload reliability."

The review was conducted by a "blue ribbon" team composed of the most capable Air Force and Aerospace personnel.[8] The conclusion reached by the audit team was that the design was not a source of "high" orbital electronic failure rate, the pedigree review was adequate to insure good flight hardware, relays accounted for 50 percent of the failures, the mechanical failure history indicated no increase since the first mission, parts were not contributors to a "high" failure rate, there was no trend of increasing anomalies, and that personnel changes were not a factor in the orbital failure rate. The review team recommended an increase in electronics test time and more severe thermal cycling.

Although the Perkin-Elmer Sensor Subsystem was designed to operate for 45 days, mission length was increased during each mission until later missions achieved over 175 days. This could only have been possible if the design and the parts were more reliable than the published reliability estimates.

The reliability philosophy followed on current space programs is essentially unchanged. Today, however, we are much more sophisticated technically and because of the availability of computers, we can now examine many designs under a variety of conditions at a faster rate and through more iterations.

MANUFACTURING AND TEST

Perkin-Elmer's basic philosophy for building the Sensor Subsystem was expressed in the Hexagon proposal. "The overall fabrication and delivery plan requires that all possible parts fabrication and some select assemblies be subcontracted. Structural assemblies and precision machining and assemblies of beryllium are typical subcontracted items. The detailed system design is being developed following this guideline. Where it has been possible, subcontractors have and will be asked to participate in finalizing of part or assembly design to insure better producibility and delivery. Fabrication of mechanical parts by Perkin-Elmer will be largely limited to airbars, parts for models or experiments, and quick response requirements."[1]

This philosophy was also applied to the fabrication and assembly of components for the electrical system. The following were subcontracted to qualified manufacturers; optical encoders, metering capstans, brushless torquer motors, and various electronic packages.

The effort facing the manufacturing and test engineers was massive. In addition to producing parts for breadboards and various experiments and tests, they were responsible for recommending and designing various test and handling equipment for the Hexagon program. They also participated in the design of the manufacturing and test facility in Danbury.

As the camera design progressed, the manufacturing engineers helped to determine tolerances to insure that the designs were economically producible. By March 1967, engineering drawings were being released for the various models which would confirm the design concepts.

The first Hexagon Camera model to be built was made of wood. It was a full-size spatial mockup that was nonfunctional

Full Size, Hexagon Camera Wood Model with Program Personnel Responsible for Design and Construction

and intended to demonstrate arrangement, major interfaces, and outlines. It was also used in the layout and design of cables and electrical harnesses.

Soon after, the Mass Model and Thermal Model began taking shape. The Mass Model was a nonfunctional assembly that was used to demonstrate arrangement, major interfaces, mass simulation, structural evaluation, and also a check on fabrication and handling procedures, mass property control procedures, and accuracy of drawings, parts list, etc. The accuracy of mass simulation for each subassembly was within at least 5 percent, with 2 percent being the design goal.[2] The Mass Model was not required to function either electrically, thermally, optically, pneumatically, or even mechanically, except that items such as the optical bar, platen, supply, and take-up were to be hand-rotatable. However, with regard to size, weight, and structural characteristics, the Mass Model closely resembled the Flight Model.

The Mass Model had to withstand qualification level shock and vibrational requirements. These tests included force-deflection and low-level vibration tests on the supply, frame, optical bar, and other camera assemblies to provide data needed to confirm the design.

Assembly of the Mass Model optical bars at Perkin-Elmer's Commerce Park Facility in Danbury began in September 1967.[3] The Mass Model two-camera assembly was completed in December 1967 and shipped to the AVCO Company for vibration testing. Assembly of the Mass Model afforded an opportunity to train Perkin-Elmer personnel and develop assembly and handling techniques for application on later models.[4] After vibration testing, the Mass Model was returned to Perkin-Elmer for evaluation. It was then disassembled and updated for testing of the frame and optical bars.[5]

The SBAC midsection was received in December 1968 and set up in the main assembly area for installation of the two-camera and the supply assemblies, cables and harnesses, and the electronic boxes.[6] After successfully passing vibration tests, the Mass Model midsection was shipped to

Mass Model

Clean Room Assembly Area

SBAC (8 April 1969) for additional testing.[7] After completion of final tests in mid-March 1970 (acoustic and pyrotechnic), the Mass Model was returned to

Delivery of SBAC Midsection in Transporter

Removal of SBAC Midsection from Transporter

▇▇▇▇▇ where it remains today in storage.[8]

Plans for the Thermal Model started at about the same time as the Mass Model. The Thermal Model was also a nonfunctioning unit and closely simulated the thermal characteristics of the Flight Model. In addition to duplicating the major component arrangement, the thermal diffusity, thermal conductance, finishes, coatings, structural mounting points and internal power distribution closely matched the Flight Configuration.

Internally mounted heaters were capable of developing 150 percent of the design heat load and the film spools were designed to carry a full load of film. Instrumentation (almost 400 thermal sensors) was located internally and externally to monitor temperature level and distribution. A detailed thermal analysis was conducted in conjunction with the design of the Thermal Model. The test data was also used in confirming the adequacy of the system thermal analysis computation methods.

By January 1968, the Thermal Model was completed and shipped to the General Electric Company for thermal tests.[9] The tests were completed two months later and the Thermal Model was disassembled. The

Thermal Model in Complete Test Fixture

parts and subassemblies were subsequently used in other tests and experiments.[10]

Fabrication of the Engineering Model began in April 1968.[11] The Engineering Model was fabricated from drawings and specifications approved at the PDR and CDR technical reviews. The purpose of the Engineering Model was to demonstrate that the system functions properly, to surface any design or manufacturing flaws, and to retrofit redesigns resulting from the

Thermal Model

Midsection Assembly Installed on Shaker Cluster

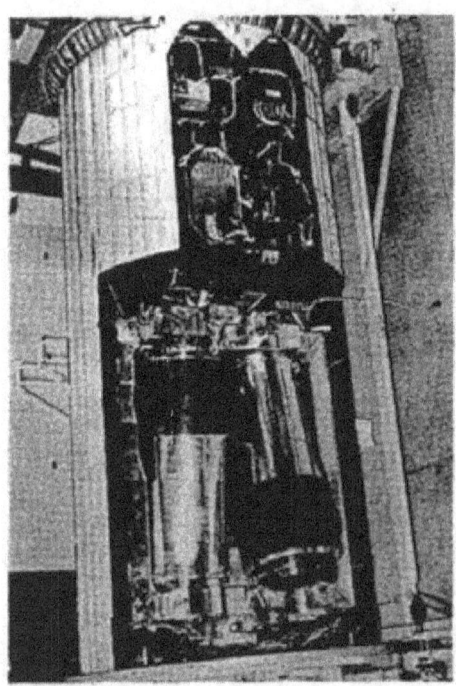

Development Model

Engineering Model tests. With the exception of a few waivers which did not compromise functional or structural integrity, the Engineering Model closely duplicated the Flight Model.

By May 1969, the Engineering Model was ready for testing of the film transport capability. The configuration consisted of a complete one-camera assembly with simulated supply and take-up. Film was transported (70 ips) with the platen in both the nonoscillating and in the oscillating mode.[12]

The following month, one-camera tests were continued in Ready Room B in preparation for Chamber B tests. The camera was operated at film speeds up to 200 inches per second at the slit, with a simulated System Command and Control

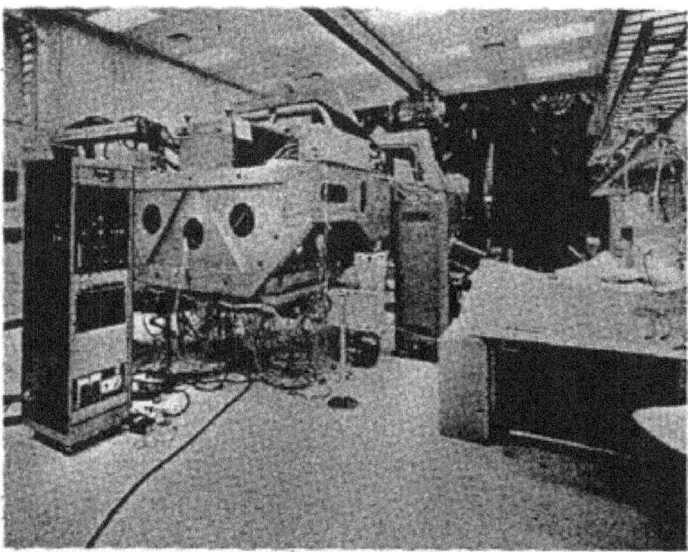
Ready Room B

and Sensor Subsystem Test Console. Testing indicated that all system parameters were being met, however, during start-up, a power supply was not turned on resulting in the platen being driven into its stops.[13]

One month later, the Engineering Model was tested in photography modes as well as the recycle mode prior to being installed in Chamber B.[14] Engineering Model tests in Chamber B were completed in August 1969. Nine photographic runs were performed both in air and in vacuum. After correcting various problems with bad film, erratic shutter operation, and high signal errors, a final run in Chamber B produced a resolution of over 150 lines in-track.[15]

By September 1969, the Engineering Model was installed into the Satellite Midsection in preparation for Ready Room A and Chamber A tests.[16] A major milestone was accomplished in the check-out of the Engineering Model in Ready Room A when the system ran in automatic mode with a preprogrammed number of frames. The Engineering Model was then moved into Chamber A on 29 November 1969 where full system operation was achieved. However, intermittent film tracking necessitated removal of the Engineering Model to determine the cause of the tracking problem.[17]

The Engineering Model was then reinstalled into Chamber A. The first series of the integrated thermal test run in a vacuum were started on 9 December 1969. The integrated thermal tests were to determine how effectively the satellite midsection maintained the temperature ($70°F \pm 23°$) of the environment surrounding the two-camera assembly, the pneumatics assembly and the supply assembly under simulated orbital conditions. Upon completion of the photographic runs, thermal conditions were changed and thermal data acquired.[18] A second series of photographic runs was started on 11 December but the run was aborted due to a malfunction of the film transport system. Unsuccessful attempts were made to free the film path. It was decided to continue the integrated thermal testing, leaving the camera system "as is."

The integrated thermal testing of the Engineering Model in Chamber A was completed in February 1970. The unit was then removed from the chamber for disassembly and post-test inspection.[20] As

Engineering Model Installation into Chamber A

Ready Room A

a result of these tests, the paint pattern and the superinsulation on the midsection were modified by the Lockheed Company. The following month, the Engineering Model was used in perfecting the techniques required to lift a system from the horizontal to the vertical position.[21]

Forward Section Simulator Mated to Midsection

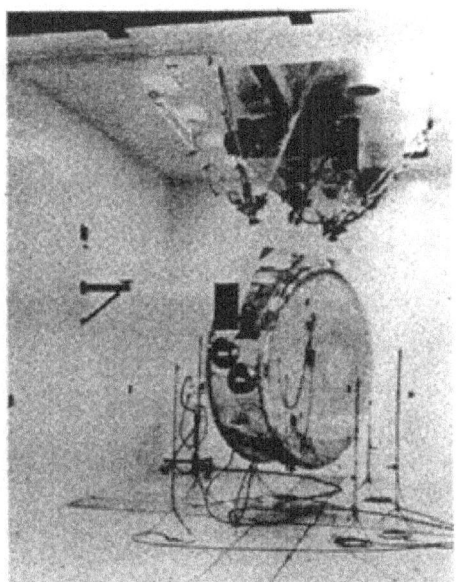

Supply Assembly Installed in Acoustic Chamber with Microphones

Chamber A

Supply Assembly Installed in Four Shaker/Fixture System

In April 1970, the Engineering Model was disassembled for inspection of all its parts prior to refurbishment for use as a film path test bed. The two-camera assembly was removed from the mid-section. The midsection was used to perfect vertical lifting techniques on the Shaker Room hoist.[22]

The Engineering Model has been in use for test purposes since the first flight mission. An "A side only" machine, it was used frequently for demonstrating the film transport system to Hexagon program visitors. Its most important function, however, was to test new film and assist in determining the causes of malfunctions during mission operation of the flight models.

Fabrication of Development Model parts began in mid-1968, soon after the approval of the CDR's. The Development model was made from Flight Model drawings and was subjected to acceptance tests and procedures used in the manufacture, assembly, and checkout of the Flight Model, including production level environmental tests.

Fabrication and assembly of the Development model was continued, and in September 1969, the Satellite Basic Assembly Midsection was delivered to Perkin-Elmer.[23] Soon after the two-camera assembly was moved to Ready

Installation of Midsection on Shaker Fixture

Room B and on 11 October 1969, the fine film path and coarse film path were spliced together and the film transport system was tested without the active drive or metering capstan. The film path was verified to have a good track.[24] By November 11, the film transport system A was successfully operated at 0.18 and 0.54 Vx/h.[25] After completion of Ready Room B tests, the two-camera assembly was removed from the simulator and installed and aligned in the midsection.[26]

Ready Room A tests followed and the

Development Model was started and a slow turn-on of Cameras A and B was completed on 7 January 1970. Difficulties were experienced with the System Command and Control Box. The unit was returned to the vendor for checkout.[27]

By February 1970, several major tests were completed on the Development Model, including the Chamber A Checkout and Photographic Qualification, Two-Camera Assembly Vibration Qualification, Horizontal Baseline and Vertical Baseline Tests.[28] After undergoing photographic thermal-vacuum qualification tests in Chamber A, the Development model was shipped to the West Coast ▓▓▓▓▓▓▓ on 10 April 1970.

In August, 1970, the Development Model, which was now assembled in the Satellite Basic Assembly, was subjected to both acceptance and qualification level acoustics vibration tests. Both the A and B sides of the Sensor Subsystem were inoperative during post-acoustics functional tests. Troubleshooting revealed that the problem was caused by lost Looper flexure screws. The problem was corrected and the system successfully passed its post-acoustics vertical baseline tests.[30]

The Horizontal Baseline Test was completed in September 1970. However, three aborts were experienced during start-up of the tests. As a result, the crossover assemblies were adjusted.[31] After acoustic, vibration, and pyroshock tests were completed, the Development Model was moved from the acoustic cell to the vertical integration stand and the shroud removed for visual inspection.[32]

By November 1970, the integrated Development Model (SDV-III) successfully completed all functional objectives of the A-1 Chamber tests.[33] Following Chamber A-2 tests, the SDV-III was prepared for shipment to the launch pad. All functional requirements were met; however, due to tracking problems caused by unclean air bars, the A-side had only partial success.[34]

The SDV-III system then completed Horizontal Functional and Vertical Pre-shipping Tests. The tests, which were limited to the B-side because of the erratic behavior of the A-side, met functional objectives. Operation of the A-side was

Sensor Subsystem Transporter

limited to cage/uncage sequences only.

The system arrived at the Vandenberg Air Force Base on 19 January 1971. After the initial operations of mating with the launch vehicle, environmental shelter verification, alignment verification, and battery installation, compatibility testing of AGE, SDV electrical, and the Automatic Data Processing and Computing System were completed.[35] The final Phase 3 tests were completed by 5 March and included an actual operation of the system.[36]

After completion of all the pad tests, the Hexagon Camera System (Development Model) was returned to ▓▓▓▓▓▓▓ on 9 April 1971.[37] A short test was run during which the Camera B side ran satisfactorily. However, the coarse film path on Side A continued mistracking due to intermittent aft steerer operation. After the tests were completed, the Development Model was temporarily stored.[38]

In June 1971, the Development Model was again placed in operation but tracking remained unsatisfactory. It was found that a film fold developed at the end of the run. Several methods of clearing the problem were tried, using the supply unit test set and cutting scallops in the film edge, with no success. Finally, the unit was removed and down loaded. It was once again noted that the A side film stack had irregularities.[39]

After running additional tests on the Development Model on the West Coast, the unit was subsequently returned to the Perkin-Elmer facility in Danbury where it was used as a vehicle for testing new film and analyzing mission anomalies.

At the same time that the various models were being assembled and tested, the polishing of the optics for the Qualification Model and Flight Models 1 and 2 was nearing completion.[40] Assembly of the optical bar for the first flight model was also started at this time.

Ready Room B testing of the Flight 1 Two-Camera Assembly was started in January 1970. Both film transports were operated in the recycle mode with bar-to-bar synchronization at $Vx/h = 0.018$ and at scan angles of 30, 60, 90, and 120 degrees

Flight Model 1 Optical Bar B

and scan center of 0 degrees. Upon completion of testing, the unit was returned to Manufacturing for retrofit. The Fight Model 1 midsection was received from the Satellite Basic Assembly Contractor at this time.[41]

The Two-Camera Assembly was completed and the camera system was tested in Ready Room B to establish a previbration electromechanical baseline for the Flight Model 1 Two-Camera Assembly.[42] A final baseline stereo run was made at $Vx/h = 0.018$ and scan angle at 120 degrees.

At the completion of the stereo run in March 1970, the Two-Camera Assembly was removed from the Satellite Basic Assembly Simulator and transported to the vibration test area. A three-axis vibration test was conducted to acceptance levels.

During these tests, the coarse and fine tension sensors in the looper and caging status of both platens and loopers were monitored. The caging systems remained caged with no failures. No significant anomalies were reported during the vibration runs. The Two-Camera Assembly was returned to the Ready Room for post vibration baseline testing.

A post vibration stereo run was again performed at $Vx/h = 0.018$ and scan length of 120 degrees. Film tracking appeared good and all gross characteristics of the

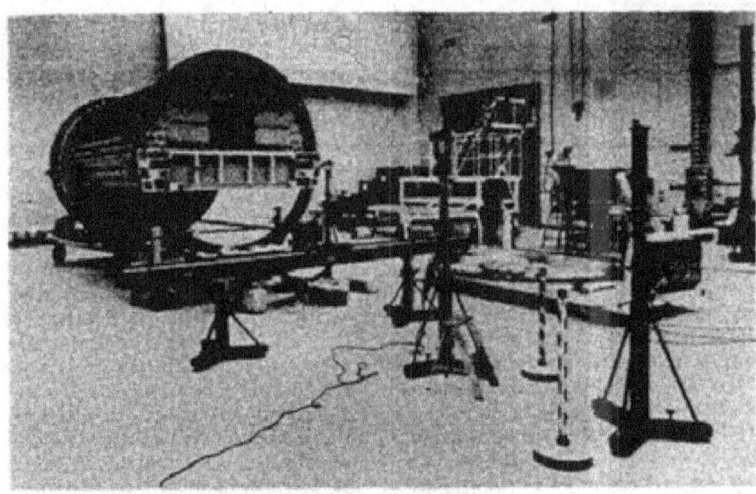

Flight Model 1 Midsection Undergoing Inspection

camera appeared to be unchanged as a result of the vibration test.

After the conclusion of the post vibration stereo run, a series of engineering tests were conducted using special edge sensors to determine the cause of the anomalous film tracking conditions apparent at high speeds and/or small scan angles.

A set of stops were made for the crossover airbars and installed. This produced a marked improvement in the tracking performance of Camera B and warranted the installation of stops in both Flight Model 1 crossover assemblies.

At the conclusion of the engineering tests, the Two-Camera Assembly was subjected to light-leak tests. Although the light-leak film was successfully threaded through Camera A, a severe film jam developed in Camera B under the slit. This was apparently caused by the thicker high sensitivity film.

Flight Model 1 Ready Room B testing was completed on 1 April 1970. The Two-Camera Assembly was returned to the major assembly area where the optical bar A encoder was replaced and realigned. The Two-Camera Assembly was then installed into the midsection with the supply and delivered to Ready Room A on 10 April. The forward section simulator was installed and system final assembly started on 12 April.[43]

On 27 April, the light leak test on the midsection was accomplished by transporting SO 380 film onto the midsection and exposing it for a period of four hours.[44] The film was spooled onto the take-up and sent to the laboratory for processing. No major problems were encountered.

The midsection was then prepared for vibration testing which was accomplished during 5 May. No major anomalies were encountered in the test. The midsection was then returned to the front of Chamber A for final testing and preparation was then initiated for thermal vacuum acceptance testing.

Flight Model 1 (S/N 002) was installed and aligned in the thermal-vacuum Chamber A at the beginning of June. The operation of the access lock, simulated take-up and the Chamber forward film path was verified. The in-air tests were completed and the stereo through focus runs were achieved using preprogrammed command tapes. The model was then removed from Chamber A to investigate a tracking problem. A design fix was implemented and the model was reinstalled in Chamber A. In-air testing was completed and preparations were made for vacuum testing.[45]

S/N 002 (Flight Model 1) began formal acceptance tests in Chamber A. The 70°F and 93°F tests were completed satisfactorily. However, an anomaly occurred in the Platen Servo Loop during the 47°F test. The Platen was electronically synchronized to the optical bar through an all digital position servo utilizing optical encoders on the optical bar and counters in the Platen electronics. Due to spurious noise, the Platen lost count and drove at maximum torque through the stops, wrapping the film and causing separation as well as severe mechanical damage to the Platen and Film Drive assemblies. This appeared to be a serious setback to the Hexagon flight schedule.

A series of emergency meetings was held with the Customer and Associate Contractors. Perkin-Elmer recommended accelerating the next sensor (S/N 003) as Flight Model 1 and agreed to attempt to complete all in-house assembly and test in three months.

At that time, a detailed plan was formulated for S/N 003 for Ready Room A and thermal-vacuum Chamber A tests. Critical test points, EMI testing, and midsection vibration testing were eliminated from the plan to improve the schedule.

During initial film tracking tests in the midsection, severe film wander was noted in both A and B film paths. Diagnostic and visual observation determined that the film was wandering at the supply film exit vestibule at a once-per-revolution of the supply. It was concluded that poor tracking was caused by a poor outer section on the film stacks. Chamber A preparations were then started for in-air testing at room temperature.

On 4 August, Flight Model 1 (now

S/N 003) was installed into Chamber A and tests started. However, a series of problems forced the Chamber A tests to be aborted on 25 August, film jams being the most serious of these problems.[47]

A post-abort investigation revealed that Camera A was jammed in the fine film path.[48] The steerer electronics were checked out and no anomalies were identified. The supply stacks were smooth and appeared normal in all respects. Several corrections were made including the installation of a new supply, new crossover assemblies of improved design, and replacement of the aft articulator assembly.

By 20 September 1970, a slow turn-on was completed and the midsection was installed in Chamber A. The very next day, tracking difficulties with the chamber film path resulted in a decision to discontinue testing with the chamber film path and to continue tests with an Engineering Model take-up in the forward section simulator. A successful turn-on, leak test and 70°F vacuum runs were accomplished by 25 September.

A major milestone was reached in October when Flight Model 1 successfully completed Chamber A acceptance testing, a series of customer-directed chamber tests, and horizontal and vertical baseline tests.[49] After acceptance by the customer, Flight Model 1 was shipped to Building 156 on the West Coast on 19 October 1970.

The schedule recovery for the first flight, by accelerating the second production sensor, is just one indication (perhaps the most remarkable) of the dedication and achievements of the Perkin-Elmer Hexagon team. It required around the clock, seven days a week activity by all concerned. In

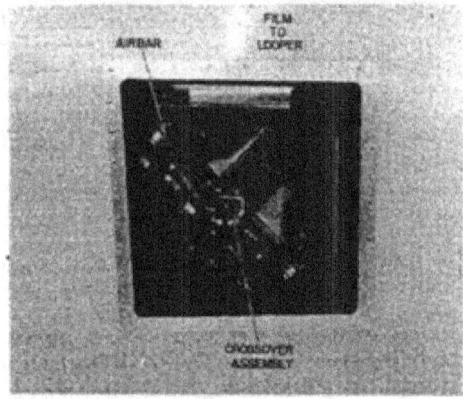

Flight Model, Crossover Assembly

Flight Model 1 Prior to Installation in Chamber A

addition, Engineering had to determine the cause of S/N 002 failure, design and implement a fix, and retrofit the sensor. All this occurred smoothly and in an almost routine manner.

The initial contract called for the first flight in December, 1970; it actually occurred in June, 1971. Perkin-Elmer's goal was to design, manufacture, assemble and test the first Hexagon sensor in 44 months with 6 months for integration and pre-launch activities. Flight Model 1 sensor was delivered in 48 months -- quite an accomplishment considering it was also necessary to construct a facility and hire many of the people, as well as design, build, and test a state-of-the-art system.

Small wonder that Hexagon is considered one of the engineering achievements of our time!

Optical Fabrication

In March 1965, Perkin-Elmer was contracted by the CIA to continue the optical design of a reconnaissance camera system started by Itek in mid-1964. After an eight-week study, Perkin-Elmer submitted a proposal (Protem) which recommended a slight change in the optical bar configuration to permit the use of a fused quartz folding mirror in place of the original beryllium mirror in the event that the latter, whose success was somewhat speculative, proved to be unacceptable.[1] In addition, the proposal included a recommendation for an investigation of the dimensional stability of beryllium and ▮▮▮▮ a relatively new material produced by the ▮▮▮▮

Lightweight mirror configurations were also a subject of a meeting held at Itek on 29 October 1964 when its engineers were faced with the problem of selecting an optical material for the "optical bar" design.[2] A major participant of that meeting, Frank Cooke, was invited to present a state-of-the-art look at the fabrication of lightweight mirrors.

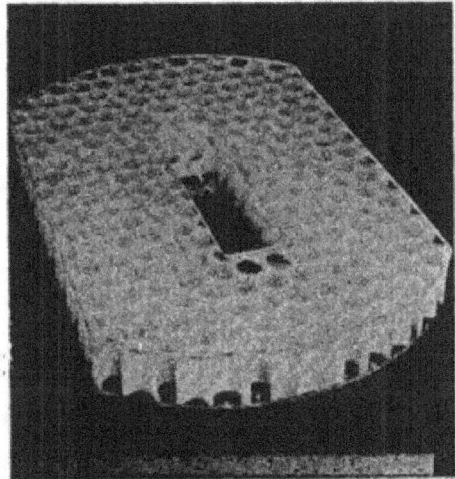

Folding Mirror Blank

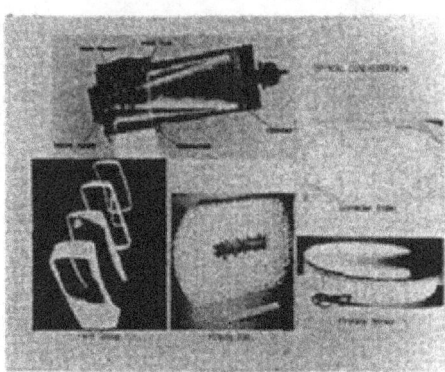

Hexagon Optical Configuration

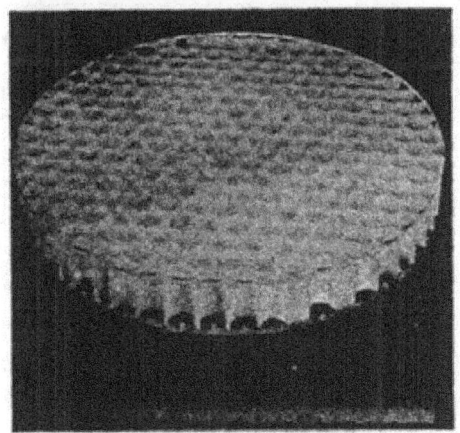

Primary Mirror Blank

Perkin-Elmer, however, had kept abreast of the state-of-the-art in lightweight mirror construction. Dr. Harry Polster, a Perkin-Elmer physicist, had previously written a summary paper on lightweight mirrors which was presented at a symposium at the Wright-Patterson Air Force Base in 1962.[3]

Before Perkin-Elmer became involved in the "optical bar" study, the CIA had already funded some optical fabrication studies at Perkin-Elmer. In March 1965, additional work was contracted (RD-2059) to conduct various studies, experiments, and analyses in the development of new optical fabrication, coating, and testing techniques.[4] These included continuous polishing, selective coating, optical test techniques, and image quality studies.

The "optical bar" design consisted of a full aperture (20-inch) aspheric corrector, an f/3 spherical mirror 26-1/2 inches in diameter, a perforated folding flat 20-1/2 inches by 30-1/2 inches, and a group of refractive corrector elements of relatively small diameter (6-3/4 to 10 inches) near the focal plane.

With the exception of the folding flat and the spherical mirror, the optics were not different in size or quantity requirements from those being routinely produced at Perkin-Elmer. Even the spherical mirror did not present any particular problems with the availability of precision test methods which were being developed at that time. The perforated flat, however, presented a significant problem due largely to its non-circular shape and the presence of a central hole.

Perkin-Elmer's Hexagon proposal noted the difficulty of producing the folding flat mirror.[5] "Normal polishing techniques for producing a flat optical surface involves reciprocating the flat on a rotating polishing lap. The reciprocating motion, which is necessary since the center of the lap polishes more slowly than the edge, produces an overturning torque on the flat which further tends to drive the

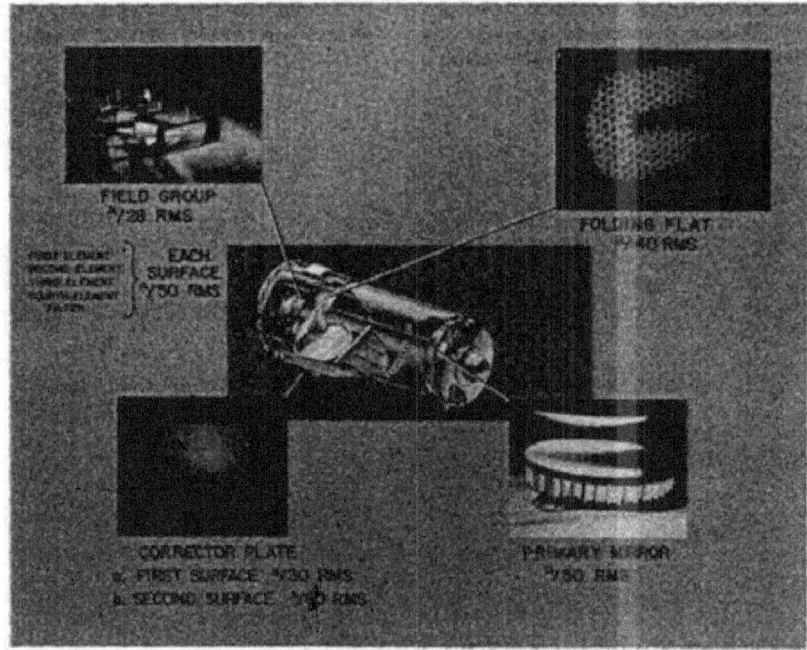

Hexagon Optical System Components

flat surface toward a sphere. A large part of the "art" in optical processing is the ability of the optician to minimize this effect, which can be done to various degrees of success for circular pieces. However, the ability to process a non-circular piece to a high accuracy is virtually impossible by this technique. Schemes such as blocking an extra glass to simulate a circular shape, differential loading, or extremely light loading, are at best compromises which generally have not produced satisfactory results." Perkin-Elmer proposed the use of fused silica of drilled core ▮▮▮▮ construction for both the diagonal (folding flat) and the primary mirror.

The Perkin-Elmer Hexagon proposal included a description of the continuous polishing technique which could produce perforated folding flats with the accuracy required. In addition, the proposal described new developments in selective coating and hologram interferometry.

By the time Perkin-Elmer was awarded the Hexagon contract (October 1966), a 48-inch continuous grinder was producing flat, fine ground surfaces of high quality, and a 48-inch polisher was producing optical flats of high quality (a 10-inch quartz disc was polished to 1/50th of a wave).[6] In addition, a 96-inch continuous grinder and a 96-inch polisher were being placed in operation at Perkin-Elmer.

After fabricating the necessary optics for the Mass, Thermal, and Engineering Models, the optical manufacturing department was ready to go into full production. Perkin-Elmer successfully produced all the optical elements for 20 flight models, in addition to supplying all the optical elements for the Hexagon Camera test equipment. This was accomplished without any significant technical or schedule problems.

The progress of the fabrication of the optical elements is reported in the Sensor Subsystem Monthly Technical Reports. A more complete and continuous picture of the development of the new optical techniques and the production of the Hexagon Camera elements can be obtained by a review of the Biweekly TWX Messages (starting with ▮▮▮▮ 2689, 27 August 1965).

4 RELATIONSHIPS AND INTERFACES WITH ASSOCIATE CONTRACTORS AND SUBCONTRACTORS

ASSOCIATE CONTRACTORS AND RESPONSIBILITIES

When the Hexagon program was in the planning stages, the aerospace/reconnaissance community included several major companies: Perkin-Elmer, Eastman Kodak, Itek, Thompson-Ramo-Woolridge, Radio Corporation of America (sensor subsystems); Lockheed Missile and Space, Martin-Marietta, Hughes Aircraft, McDonnell-Douglas and General Electric (spacecraft); and General Electric and AVCO (reentry vehicles). Eastman Kodak, of course, supplied all the reconnaissance film.

The CIA's initial contract (1964-1965) for the fourth generation reconnaissance system (then codenamed Fulcrum) was awarded to four companies: Itek was selected to design and produce the camera payload, General Electric was to design the spacecraft, and the AVCO was awarded the reentry vehicle contract and TRW was the System Engineering and Assembly contractor.[1]

It was during this time frame that the government policy decision to assign the roles of the Air Force and the CIA in the reconnaissance activity was being formulated. After Itek withdrew from the CIA Fulcrum Program in February 1965, they worked on the Air Force version of the fourth generation reconnaissance system code named S-2. The CIA selected Perkin-Elmer to continue the camera design started by Itek.

The government policy decision to assign the spacecraft to the Air Force, and the reconnaissance sensor payloads to the CIA, negated the CIA contracts awarded to General Electric, and AVCO.

This brought all the players back to square one. A competition for the new reconnaissance system (now codenamed Hexagon) was established and a Request for Proposal was released to selected companies.

On 10 October 1966, Perkin-Elmer was awarded the camera payload contract, Lockheed won the spacecraft contract, and the reentry vehicle contract was awarded to McDonnell Douglas. These were the major participants in what was to be the most complex reconnaissance system ever envisioned.

SELECTION OF SUBCONTRACTORS

Perkin-Elmer's procurement planning for the new reconnaissance system began in September 1964, when it got involved in Phase I of the Fulcrum program. In May 1965, a technical report was prepared identifying subcontractors and vendors suited to Fulcrum program requirements. It listed 20 technical consultants and over 100 vendors.[1]

By the time the Hexagon Request for Proposal was sent to Perkin-Elmer in May 1966, procurement policies had already been established for the program, and vendors that would participate had already been contacted and surveyed. The major subcontractors identified in the Perkin-Elmer proposal included the ▓▓▓▓ Pennsylvania, ▓▓▓▓ Alabama, and ▓▓▓▓ Massachusetts.[6]

Initially, the Purchasing Department not only processed the requisitions for components and small machined parts, but was also involved in the selection of major subcontractors for the larger structures and assemblies. Later, as the program became organized, a Subcontracts Department was created. It was decided that the Purchasing Department would process orders for fixed price items, and the Subcontracting Department would handle the larger parts requiring other type contracts. The dollar value of the item was also a consideration. The exception to this arrangement was the raw glass, purchased on a fixed price basis, which was

145

handled by the Subcontract Department.[3]

One of the most difficult problems confronted by the customer and Perkin-Elmer was the processing of millions of dollars of purchase requisitions and subcontract work without divulging the purpose or function of the purchased parts. To assist in this process, Perkin-Elmer formed a dummy corporation, JETEC. Invoices from and payments to the major subcontractors were "laundered" through this dummy corporation to conceal the amount of work involved between the various companies and Perkin-Elmer.

It was necessary, in some cases, to provide program access to some of the top managers of the various vendors and subcontractors, but this was held to a minimum. A security evaluation of subcontractors was prepared in September 1967 listing ten companies. When it was necessary to establish secure areas in subcontractor's facilities, Perkin-Elmer security officers determined if the secure areas and procedures were sufficient to prevent compromise of program knowledge.[4]

By December 1966, just two months after the award of contract, Perkin-Elmer vendors and subcontractors were engaged in preliminary studies and fabrication of parts for the various Sensor Subsystem models. One year later, the Perkin-Elmer Subcontractor team was in the initial stages of fabricating and assembling parts for the production models.

Fabrication of Optical Bar

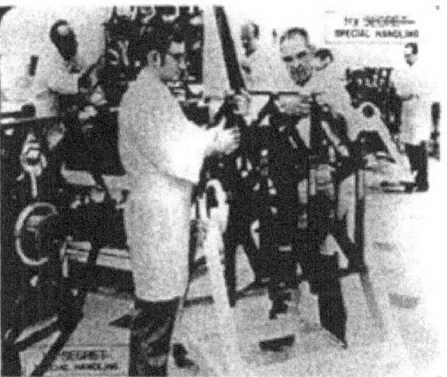
Take-Up Assembly

Fabrication of Frame

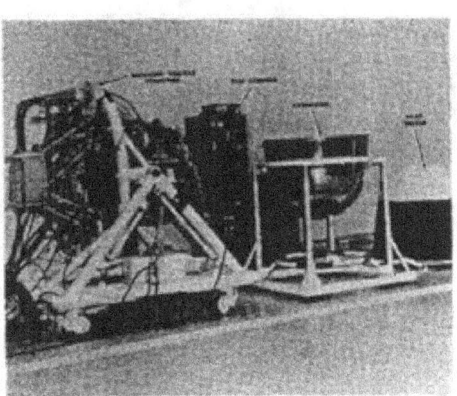
Test Set-up for Take-Up Assembly

A monthly status report was prepared by the Subcontracts Department. The June 1968 report listed the following companies: ███████████ (later called ███████), Radio Corporation of America, ███████████████████ Lockheed Missile and Space, ███████.[5]

A general understanding of Perkin-Elmer's initial involvement with vendors and subcontractors can be gained by the following excerpt of an interview with one of the early purchasing agents on the Hexagon program.[6]

"Initially we were involved in getting a lot of breadboard hardware and preliminary quotations and backup material. I remember, in particular, a requirement for fiberglass replicas of the take-up containers. They were spinnings from hand-laid fiberglass cans. We also purchased breadboards of the Optical Bar tube structures before we got into actual hardware.

We used local machine shops. The Optical Bar tubes were made by ███████████████████████████. They also did a lot of initial breadboards. For machining we used ███████████████████, etc. For sheetmetal fabrication we used ███████████████. Electronic parts were purchased from Command and we also used catalogs, mostly distributor stuff... that was before the need for high reliability parts.

We worked a lot with hand sketches in the early days. The prime example are the vacuum chambers. The original concept for this equipment was on 8 x 11 sketches. An overnight trip was made to Pittsburgh and also ███████████. We got an estimate in two days based on those sketches... I believe it was ███████ dollars for the whole thing. We did similar things for the frame and the optical bars tubes and other major structures.

After the award of contract to Perkin-Elmer, preliminary drawings were released and we selected the most qualified sources based on their response to our request for quotes. A good number of the original key vendors still exist and are still involved on the Hexagon program."

DEVELOPMENT OF INTERFACES

The techniques used to manage interfaces on major aerospace programs were fairly well established by the beginning of the Fulcrum program (Phase I) in 1964. At the time a government agency released a Request for Proposal for aerospace equipment, it required that competing companies include an Interface Requirements Section in their proposal. The agency later used this information to develop an Interface Requirements Document (IRD) which it included in the final contract.

This document was used as a basis for establishing interfaces between associate contractors on the program. Agreements reached by interfacing contractors were documented in an Interface Control Document (ICD). The ICD is essentially an agreement between associate contractors and forms the basis for the responsibilities and actions of the two companies.

Interface Working Group (IFWG) Meetings, which covered specific disciplines (i.e. electrical, thermal, etc.) were held between associate contractors. During these meetings, the contractors reviewed, line-by-line, the contents of the particular ICD being discussed. After both parties were satisfied, the ICD was then signed.

When Perkin-Elmer first became involved in the Fulcrum program (September 1964) it collected all interface information provided by the customer in a System Specification Book.[1] This book was later used as a basis for the AD HOC Specification Book which was started after the customer switched their reconnaissance program from Itek to Perkin-Elmer in March 1965.[2]

Soon after, the customer sent a TWX message to the spacecraft and re-entry vehicle contractors (General Electric and AVCO) instructing them to convey to Perkin-Elmer complete details of all aspects of the Fulcrum Program.[3] A meeting of these companies and Perkin-Elmer was held on 1 April 1965 to discuss interfaces. Additional interface meetings were held later in 1965, however, these meetings were for information purpose only and did not form the basis for any ICD's.[4,5]

During a Perkin-Elmer technical review meeting with the customer in December 1965, an organization chart of the new division (which would be established if Perkin-Elmer was successful in winning the Hexagon Sensor Subsystem contract) was presented.[6] It included an Interface and Liaison Control Group which would report directly to the Hexagon Sensor Subsystem Program Manager at Perkin-Elmer. This was the nucleus of what later developed into the Hexagon Program Interface Group. During a reorganization in early 1967, the Interface Group became a part of the Systems Engineering Department.

The Request for Proposal issued by the government in May 1966 for the Hexagon Sensor Subsystem contract included the following references to interfaces.[7] "The Contractor shall perform analyses and studies necessary to provide the Procuring Agency with detailed functional and physical interface requirements and constraints to be included in the definition of interfaces between the Sensor Subsystem and the Satellite Basic Assembly, Recovery Vehicles, Stellar/Index Camera, and Space Vehicle AGE/facilities.

After the approved interface documentation has been contractually implemented as part of the Sensor Subsystem Performance Specification, the Contractor shall ensure that the design, as it evolves, complies with the interface requirements. Design changes which affect the interface shall be submitted to the Procuring Agency for approval.

The Contractor shall support the Procuring Agency in interface meetings with other Agencies and Contractors as required to negotiate interface changes proposed by either side and to resolve other interface problems as they arise. In this context, the Contractor shall assist the Procuring Agency in evaluating interface changes proposed by other agencies with particular emphasis upon the impact of the change on the Sensor Subsystem performance and design, test program, delivery, schedule, and cost."

Prior to the award of contract (10 October 1966) the Perkin-Elmer Interface Group issued a preliminary Interface and Liaison Program Plan which identified its functions and plans for conducting interfaces activities on the Hexagon program. This was later used as the basis for the Interface Management Manual and Interface Control Procedures Document.[10] The group had previously prepared the interface requirements which were included in Perkin-Elmer's Hexagon Proposal to the customer in July 1966.[11]

On 26 October 1966, George R. Gray, who headed the Interface Group until 1977, released a "kick-off" memorandum establishing preliminary interface meetings.[12] However, he would be working under a handicap since the satellite vehicle and re-entry vehicle contractors had not been selected. (The General Electric and AVCO contracts on the Fulcrum program had been canceled by this time).

The first SSC/SBAC (Perkin-Elmer/Lockheed) informal discussion was held on 30 August 1967 to acquaint Lockheed with the general arrangement of the Sensor Subsystem and to obtain information on Lockheed's concepts in various mechanical and structural interfaces on their satellite vehicle.[13]

This interface activity was reported in Perkin-Elmer's Monthly Technical Report to the customer (September 1967). This was the first Interface Group input to this report. Monthly Interface reports continued until July 1969.

The customer released the Interface Requirements Document (IRD 501) in March 1967.[14] The document identified Sensor Subsystem interface requirements with respect to other associate contractors which included Lockheed (Satellite Basic Assembly), McDonnell-Douglas (Recovery Vehicle), and Itek (Stellar/Index Camera).

Two basic categories are covered in aerospace interface documentation; (1) interfaces of the flight hardware and (2) interfaces of assembly and verification, which includes interfaces of ground support equipment, special test equipment, and the manufacturing and test facilities. Within this grouping there exists a matrix of interfaces divided into four four disciplines; Structural, Mechanical, Electrical, and Thermal.

As previously mentioned, the IRD is used as a basis for discussion between contractors to produce an Interface Control Document (ICD) which covers a particular area (e.g., Electrical). The IRD 501 would be used as the controlling document until the customer and Perkin-Elmer agreed that the ICD's adequately identified all the interfaces. Particular paragraphs in IRD 501 would be retired as ICD's were approved.

The Perkin-Elmer Interface Group worked with the associate contractors to produce ICD's in a timely manner in support of the program schedule, but at the same time was careful that the documents were as complete and accurate as possible before being released. Changes to an ICD were very costly and would have impacted the schedule.

The first SS/RV (Perkin-Elmer/McDonnell-Douglas) ICD discussion was held in June 1968. The SS/RV ICD was signed on 4 October 1968. By March 1969, most of the ICD's on the Hexagon program were approved. A Sensor Subsystem Critical Design Review, held that month, listed 37 SS/SBA and 12 SS/RV ICD's.[15]

From the beginning of the program in 1964, SETS (customer technical consultant) acted as their interface manager. On 9 July 1968, Perkin-Elmer was informed that full responsibility for interface management on the program would be transferred to Perkin-Elmer with Arnold Wallace designated as the SSPO Interface Manager.[16] The customer memorandum indicated that a substantial part of the interface activity was completed and noted, "Although there has been considerable recent interface activity, this will not terminate interface work. It is anticipated that the number of unresolved interface areas, and the very large number of numerical ICD values marked, 'To Be Determined', together with upcoming RV interface problems and others that develop as designs become more complete, will require interface activity to continue at a moderate high level for at least six months and taper off somewhat thereafter."

During an interview, George Gray noted that the cooperation between the associate contractors was responsible for the effective manner in which the interface meetings were held.[17] He attributed this to the people that had been selected to work in this area and stated, "The end product of all interface activities and meetings is the signed ICD. Since this phase of engineering deals with information that becomes legally binding, the documents must be clear and concise and not subject to various interpretations. Personnel in this area must be good negotiators since they must deal not only with engineering personnel in their own company, but also with representatives from other companies."

The success of the ICD's and the working group was demonstrated by the fact that the first flight unit was integrated with only one minor glitch on one wire.

THIS PAGE INTENTIONALLY LEFT BLANK

5 SYSTEM INTEGRATION, LAUNCH, ORBITAL OPERATIONS, AND RECOVERY

DEVELOPMENT OF THE WEST COAST FIELD OFFICE

The West Coast Field Organization, which supported the Hexagon program, had its beginnings in the fall of 1966.[1] The group was formed as part of an overall Test Department and was located at 77 Danbury Road, Wilton, Connecticut. The initial responsibility of this group was to develop a preliminary plan for the test, evaluation, and operation of the Hexagon Sensor Subsystem.

In November 1967, Michael Maguire, who at that time was Director of Operations on the Hexagon program at Perkin-Elmer, hired Charles O. Bryant as the Department manager of Field Operations. Bryant, who previously worked at the General Electric Company when it had the satellite vehicle contract on the Fulcrum program, had many years of experience in the reconnaissance community.[2] Bryant realized that for Field Operations activity to be effective, it had to report directly to the Director of Operations and accepted the position on the condition that Field Operations was elevated to department level.

After the reorganization, the Field Operations Department began to develop the flow diagrams for the field testing of the Sensor Subsystem, the interface control documents, and equipment requirements. Many liaison and coordination trips were made by Field Operations personnel between the East Coast and the West Coast to establish a working arrangement with the associate contractors on the program; Lockheed and McDonnell Douglas.[3]

Lockheed was in the process of building a new facility for the assembly and checkout of the Hexagon System. Perkin-Elmer requirements were incorporated into the building layout. The Lockheed facility, located at Moffett Field, Sunnyvale, California was known as ▓▓▓▓▓▓.

Meanwhile, the new Perkin-Elmer Danbury facility was completed at Wooster Heights and the Field Operations activity was moved in March 1968. Additional people joined the Field Operations Department at that time. William Cottrell, an engineer who previously worked for the CIA on another reconnaissance program and had security access to the Hexagon program, joined Perkin-Elmer as manager of the West Coast office.

Departure of Bill Cottrell (left) from Perkin-Elmer. Frank Harrigan, Jr. (center) and John McNerney

The first Field Operations employee on the Hexagon program who was sent to the West Coast on a permanent transfer was Harry Loper, Administrator of the Field Operations Department. He leased and furnished a "white" office on Saratoga Avenue in San Jose, California which became the base of operations for the Field Group for almost two years. This office was later moved to Santa Clara 22 July 1968. The office remained in operation until 1 July 1971, when the contract for the "white" business office expired. The "white" Field Operations Office was moved to a location within a building occupied by the Ultek Division of Perkin-

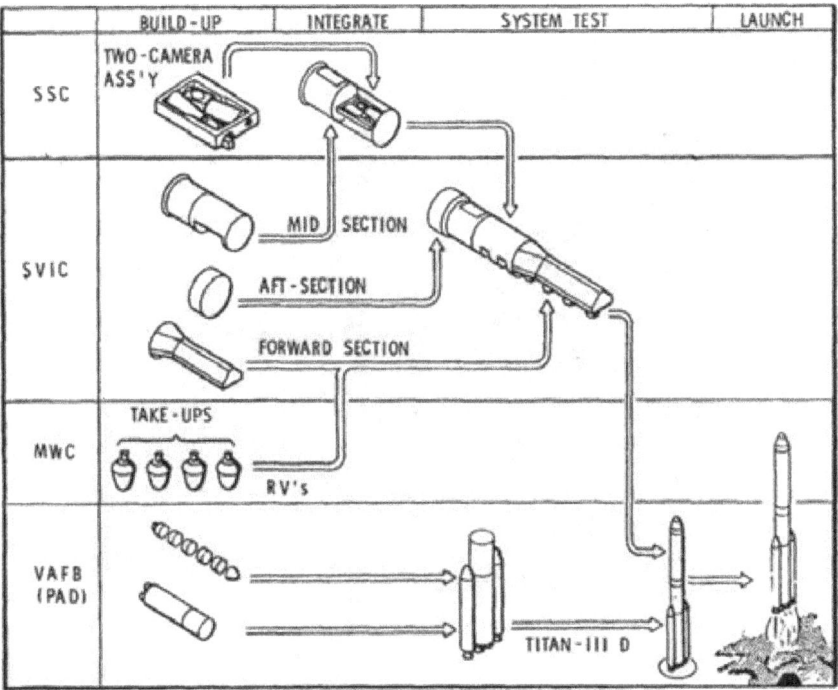

Factory to Pad Build-Up Sequence

Elmer in Mountain View, California.[4,5]

Bryant, Cottrell, and the small nucleus of Field Operations personnel began to develop the integration and test procedures that would be required to assemble the Sensor Subsystem and the reentry vehicles to the satellite vehicle on the West Coast, and conduct all the tests necessary prior to the delivery of the Satellite Vehicle and payload to the launch pad at the Vandenberg Air Force Base.

In addition to managing the Field Operations activity, Bryant and Cottrell interviewed and hired over 100 skilled engineers and technicians. A large number of these hires were people acquired from two companies. The MOL program had been canceled and a group of people from General Electric were hired from that program. Another large group came from Hiller aircraft when their operation phased out.

In February 1969, a small group of Field Operations personnel moved into temporary ("black") offices in and by November of that year, the entire organization (except for East Coast representatives) was moved to its permanent location in the Lockheed Building.[6,7]

Soon after, test equipment and test stations began to arrive from the various subcontractors and Perkin-Elmer, including the RV test set, the electrical simulator, and numerous handling devices and equipment.

In the meantime, staffing of the field office at McDonnell-Douglas was planned to begin in May 1969. This office was established to support the assembly and testing of the take-ups in the reentry vehicles being manufactured by McDonnell Douglas.[8]

During the time that the Sensor Subsystem was being fabricated, assembled, and tested at Perkin-Elmer in Danbury, West Coast Field Operations

Space Launch Complex-4 East (SLC-4E)

(WCFO) engineers and technicians periodically visited the facility to assist in testing the Development and Flight Models. In addition to providing assistance to the production program, WCFO personnel received practical training in handling and operating the Sensor Subsystem before the models were shipped to ▓▓▓▓▓▓.[9]

The first unit shipped to the West Coast was the Mass Model. It arrived at ▓▓▓▓▓▓ in 8 April 1969. Final tests on this unit were completed in mid-March 1969. Almost one year to the day, on 10 April 1970, the Development Model arrived at ▓▓▓▓▓▓. This model would enable the WCFO personnel to put the assembly and test facility through its paces — the initial "shakedown cruise".

The Development Model (SDV-3) was processed through the tests as if it were a flight model. The purpose was to verify all the test interfaces and procedures. WCFO personnel worked 12 hours a day, seven days a week. It required over eight months to complete all the tests and exercise the facility and the equipment. Finally, the Development model was shipped to the Vandenberg launch pad on 19 January 1971 and remained there for three months undergoing various tests and rehearsals.

FINAL ASSEMBLY AND TESTING OF FLIGHT MODEL 1 (SV-1)

On 19 October 1970, Sensor Subsystem, Flight Model 1 was transported from the Perkin-Elmer Danbury facility to the Bradley International Airport, Hartford, Connecticut and loaded into a C133B cargo military aircraft.[10] (Prior to December 1969, all air shipments departed from the Stewart Air Force Base, Newburgh, New York.)[11,12] Flight Model 1 arrived safely at the Naval Air Base, Moffett Field, Sunnyvale, California and was transported to ▓▓▓▓▓▓ which is ▓▓▓▓▓▓.

Following a physical inspection and optical alignment of Flight Model 1, the take-up simulator was installed and preparations for the Receiving and Inspection (R&I) runs were completed. A slow turn-on was initiated on 24 October, however, miscellaneous problems

Loading of C-133 Military Transport

Sensor Subsystem Received at West Coast

delayed the start of the R&I test sequence until 26 October.

The first series of tests in Block I of the System Command and Control were completed without incident, but transfer to System Command and Control Block II was unsuccessful. It turned out to be a command sequence problem and not a Sensor Subsystem problem. However, as a result of command sequence errors, the system went into an emergency shutdown condition, resulting in a film jam (the first on a Flight Model at the West Coast). On 30 October, the R&I test sequence was again underway.[3]

On 3 November, the supply was removed for reloading. The supply was also reworked to correct some known minor discrepancies and to retrofit the repressurization valve. While it was on the hoist, the supply was struck by SBAC's dolly tractor, causing a small dent in the supply cover. Post loading tests, including the pressure and caging tests, were rerun

Final Assembly of Hexagon Satellite

satisfactorily. It was decided to reinstall the supply assembly into the midsection "as is."14

System verification sequences were performed and approximately 50 percent of the R&I sequences were completed by 10 November. At that time, the tests were halted to retrofit the Looper Assembly and replace other components with improved designs. Upon completion of the retrofit, film path pressurization and tension were reestablished.

On 18 November, the Sensor Subsystem (SS) creep in manual SSTC at 5 inches per second was completed with no problems. At the start of the fourth run, the system shifted into maximum rewind. The system was powered down, at which time it accelerated to maximum forward, resulting in a film jam. This resulted in a change to the system operating procedure.

The supply was removed from the midsection on 20 November. Several wraps in reverse position were noted on the film spool, along with loose material on the bottom of the supply. The supply was down-loaded for repair of the brake pads, caging straps, and verification of both A and B drives. The supply was reinstalled into the midsection on 25 November for the completion of R&I testing.

While Flight Model I was being tested in ▓▓▓▓▓▓ the first in-orbit rehearsal was in progress at the Satellite Test Center (the Blue Cube). These rehearsals continued from 9 November 1970 to April 1971 (total of five).5

After R&I tests and final changes were completed on the midsection, Flight Model I was turned over to SBAC (Satellite Basic Assembly Contractor) for mating to the SV-1 Forward Section and the Satellite Vehicle, on 4 December 1970. The mating progressed smoothly.6

SBAC's aft section testing, using the Sensor Subsystem electrical simulator, continued until 20 December, during which time a number of problems were encountered requiring electrical box replacement. A creep test and a vertical baseline test were completed with no anomalies.

SBAC continued troubleshooting and module testing until 31 December, when the vehicle was installed in the horizontal

SV-1 Mated in Vertical Integration Stand (Shroud is Shown on Right-Hand Side)

SV-1 in Acoustics Chamber

test stand.

All functional objectives were met when SV-1 completed the horizontal baseline and mission profile tests. During acoustic preparations, the thermal blankets were installed, the cables were tied, and the alignment mirrors on the articulator were removed.[7]

The pre-acoustic vertical baseline was successfully completed on 21 January 1971 followed by the acoustics environmental test on 22 January. The system was then moved to the vertical test stand for post acoustics functional test. Inspection of the Forward Section was completed with no visible discrepancies. The post acoustics vertical baseline test was successfully completed with no anomalies.

Acoustic vibration exposure and post acoustic tests in the vertical integration stand were completed on 28 January. All test objectives were met. During Chamber A-1 (thermal-vacuum) preparation, several retrofits were performed.[8] However, there was a change in plans, and Chamber A-2 tests were scheduled before Chamber A-1 tests.

Chamber A-2 preparations were completed and the SV-1 was installed in Chamber A-2 on 17 February. The in-air tests were completed and the film was retrieved on 18 February. A builder roller problem was discovered on A-side of RV-1 and a decision was made to transfer to RV-2 and continue Chamber A-2 tests. The chamber reached temperature stabilization, confidence runs were completed, and the test sequence was started on 24 February.

By the end of March, Chamber A-2 photographic tests were completed. SV-1 was installed and instrumented in Chamber A-1 and thermal tests were completed with no real time anomalies. The chamber was then repressurized in preparation for removal of SV-1.[9]

The film was then retrieved from all four recovery vehicles. A film foldover was noted in the RV-1 take-up (A-side). The foldover occurred about halfway through the inadvertent 47-minute run and corrected itself without causing an emergency shutdown. Film wedging was determined to be the cause of the problem.[10]

Horizontal preshipment tests were successfully completed and R-6 day of launch countdown were completed during the vertical preshipment tests. At that time, a decision was made by the customer to replace the slit and shutter on both platens of SV-1. The system was placed in a horizontal position and the crossover assemblies, film drives, and platens were removed. The platen assemblies were returned to the Perkin-Elmer Danbury facility on the East Coast.

The platens were reworked and reinstalled in SV-1. Tracking tests were started; however, the A-side platen did not

Sensor Subsystem in the Horizontal Holding Fixture (Tilt Dolly)

Installation of Sensor Subsystem into the A-2 Vacuum Chamber

function properly. Film drive A was removed and the platen caging mechanism was discovered to be damaged. Since it was no longer used, the mechanism was removed and the film drive reassembled. Tracking tests were performed and the film was recovered.

An abbreviated Chamber A-2 test was completed with no real time anomalies. SV-1 was removed from the chamber and the film was recovered. Horizontal preshipment preparations were completed successfully and SV-1 was moved to the vertical integration stand. Abbreviated vertical preshipment tests were then run.[11]

On 4 June 1971, Shipping Certification for SV-1 was approved certifying that, "SV-1 has satisfactorily completed all required testing and flight preparations at the SBAC facility and is ready to ship to the Vandenberg Air Force Base (launch pad SLC-4E) for final flight preparations."[12] The Sensor Subsystem was caged and the SV-1 was transported to the launch pad. Prelaunch sequences were accomplished without incident and a flight readiness meeting concluded that SV-1 was ready for launch.[13]

On June 14, 1971, ███████, signed the Launch Certification Document for SV-1 indicating that "SV-1 has satisfactorily completed all required testing and flight preparations at VAFB and is ready to continue into the final countdown and upon successful completion of the countdown to launch."[14]

MISSION ACTIVITIES

SV-1 was launched into a near-perfect polar orbit on 15 June 1971. Camera diagnostics were nominal immediately after launch and the camera became operational on 16 June. By 20 June, approximately 20,000 feet of film had been accumulated on each side of the RV-1 take-up.[1] RV-1 was separated from the satellite and made a normal entry into the earth's atmosphere. However, because the parachute was damaged, no attempt was made for air recovery. The capsule was

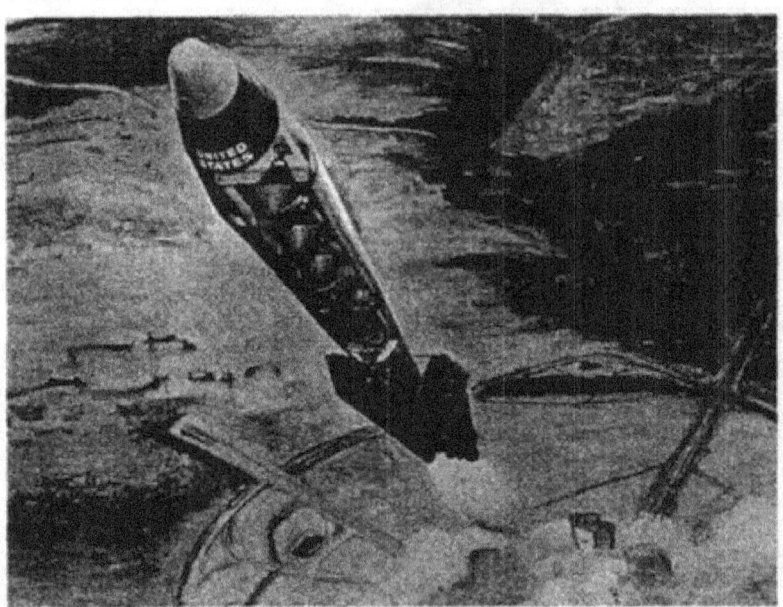

Launch of the Hexagon Satellite Vehicle

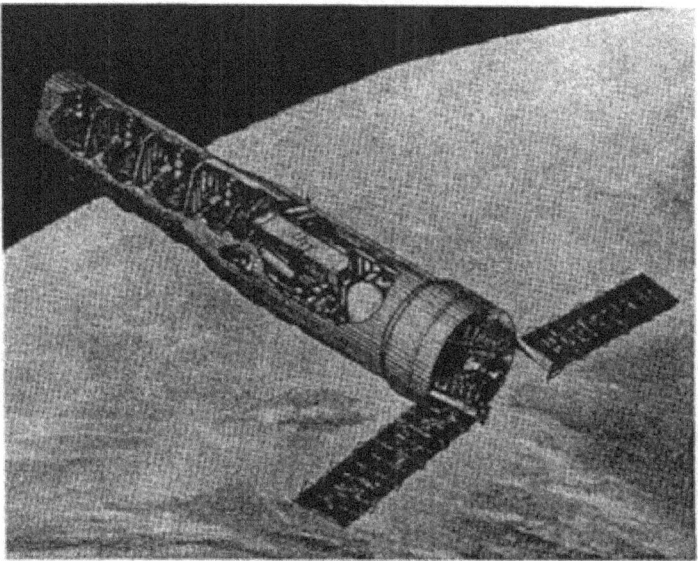
Hexagon Satellite Vehicle

allowed to land in the water where it was retrieved by scuba divers and a helicopter. RV-1 arrived at the despooling facility on 21 June and processing was completed the following day. After film despooling, the take-up assembly was removed from the RV and tested. These tests indicated no obvious deterioration due to launch, orbital operation, or recovery.

RV-2 was recovered on 26 June after accumulating 26,100 feet on each side of the take-up.

Operation of RV-3 was completely routine and it was loaded with approximately 27,000 feet of film on each side.[2] On 10 July 1971, RV-3 was lost in the water when the main parachute failed to open. Due to the chute problem, limitations were placed on the amount of film to be loaded into RV-4 which was recovered in air on 16 July with 13,000 feet of film on each side.

Four emergency shut-downs (ESD's) occurred during the SV-1 mission. The first occurred on 5 July (revolution 314) when the B-side looper struck the stops. A constant velocity run cleared this problem.[3]

Again on 10 July (revolution 402), a B-side ESD occurred due to loss of film tension at the take-up assembly. A constant velocity run restored tension, and the system was returned to operation.

On 13 July (revolution 450), the third ESD occurred, apparently due to a film jam in the B-side film path. A series of "mini-creeps" failed to clear the problem. A plan for operating the system in reverse was proposed but decided against. The film was "jerked", but the problem persisted and was diagnosed as a stuck drive capstan. A recycle was recommended as the only safe way to reverse film direction at the drive capstan. The recycle was executed and the problem cleared. Operational photography resumed.

The fourth ESD occurred on revolution 492 due to a command decoder problem. The system resumed immediate operation.

After separation of RV-4, the "solo" phase (engineering tests) of the mission

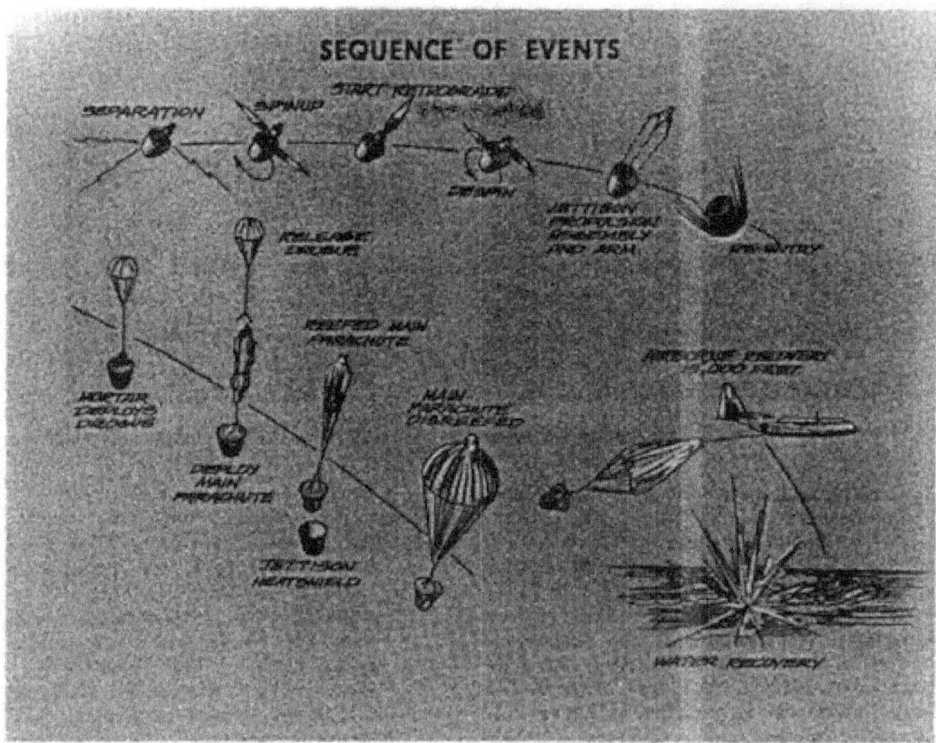

Separation of RV's from the Hexagon Satellite Vehicle

was initiated. Sensor Subsystem participation in this phase included optical bar bearing tests, operating through SSC-2, slit width tests, and focal plane position tests. "Solo" tests continued until 6 August 1971 when the vehicle was returned to the earth's atmosphere and destroyed.

RECOVERY OF THE LOST RV-3

On 27 July 1971, even as SV-1 was still in orbit, a planning meeting was held to recover RV-3 which was reported to be in the Pacific Ocean approximately 360 miles north-northwest of Pearl Harbor, Hawaii at a depth of 14,400 feet. Attending the preliminary meeting were representatives from the customer's office, Perkin-Elmer, McDonnell Douglas, Eastman Kodak, and the U.S. Navy.[1]

The Navy proposed the use of the deep submergence vehicle, Trieste II, which at that time was certified to a depth of 13,000 feet. However, the Navy was confident that this depth could be exceeded without danger to the vehicle or the crew.[2]

Additional working sessions were subsequently held to assess the probable damage to the RV on impact, define the configuration of the payload on the bottom of the ocean, estimate the drift due to ocean currents, develop techniques for attaching the payload to the lifting cable on the recovery vehicle, Trieste II, design the hardware and establish the interfaces required to perform the recovery operation.[3,4,5]

Perkin-Elmer was assigned to design the recovery hook and the Navy agreed to fabricate it by 16 September. A model of

The Trieste, A Deep Submergence Vehicle

Recovery Hook in Locked Position

the "Hay" hook was built by Perkin-Elmer and demonstrated.[6]

The initial schedule for the recovery of the RV-3 capsule was very optimistic. According to a memo dated 27 July 1971, "The Trieste will recover the vehicle sometime in September. The despooling operation should plan on starting about 27 September 1971."[7] The chronology of the actual operation as compiled by ▓▓▓▓ follows.

Sep. 20, 1971 - The Trieste is being outfitted on a dock at the San Diego Naval Base (Submarine Development Group I). Earlier this morning, hook tests were performed on a beach at the base using a dummy RV. Prior to leaving for the recovery site off Hawaii, a test dive of the Trieste will be conducted off the California coast. It is planned to rehearse the recovery of a "test RV."[8,9]

Sep. 21 - The rigging operation in preparation for loading the Trieste aboard the support ship, the White Sands, started this morning. The shipping container in which the RV-3 is to be loaded immediately after recovery is already on board. Since the recovered RV has to be maintained at a temperature below 40°F to prevent fungus growth, it is planned to place the shipping container into a wooden box filled with dry ice.[10]

Sep. 23 - Preparation of the White Sands and the Trieste is being completed. Since a daylight recovery is now planned, it is necessary to fabricate a cover for the payload made of black nylon cloth. The cover will be attached to the hook and will completely cover the hook and the recovery vehicle. It has "draw strings" that will be pulled by the divers and closed at the top and bottom while the hook and RV are still in the water at a depth of 35-40 feet. Meanwhile, the Navy is planning to construct a refrigerated wooden box to enclose the RV shipping container while in transit to the recovery site near Hawaii.[12]

Sep. 24 - Plans to depart this day from San Diego for the recovery test site off the California coast are delayed until Sep. 27 due to equipment problems on the White Sands. Because of this delay, only one test dive is now planned. A meeting was held aboard the White Sands to discuss the techniques which will be used to direct the Trieste to the target at the bottom of the ocean. Attending the meeting were representatives from ▓▓▓▓ Perkin-Elmer, the Scripps Oceanographic Institute,

Trieste Being Towed by Small Boat to Floating Crane

Trieste Being Lifted Aboard White Sands

Apache, A Seagoing Tug

and the three-man crew of the Trieste. The plan is to proceed to the test range west off San Diego which is instrumented in a manner similar to the network at the Hawaiian recovery site. The test RV is equipped with a "pinger" and will be dropped over the side. The Trieste will make one dive (about 5000 feet), retrieve the payload, make the transfer to the White Sands, and be brought aboard the White Sands. The ship will then head for the Hawaiian recovery site.

Sep. 27 - The White Sands left the dock at 3:30 P.M. and is being towed by a sea-going tug, the Apache. A meeting aboard the ship reviewed the various search, navigation, and homing techniques. In general, the plan is to lay a network of transponders on the ocean floor, survey them using satellite navigation and once the payload is located, to reference the payload location to the transponder and mark its location with a pinger. There is no automatic equipment on either the Trieste or the surface ships to determine, with any degree of accuracy, the location of the Trieste under the surface. A scheme using hydrophones and transponders will be used to direct the Trieste to the test RV.[13]

Trieste Floated Out of Dockwell of White Sands

Trieste Being Prepared for Test Dive

First Trieste Test Dive Off California:

Sep. 28 – The plan is to unload the Trieste from the dock of the White Sands and fill its tanks with gasoline and ballast shot. In summary, the plan is to carry the test RV to the bottom on the Trieste and cut it loose. The Trieste will then maneuver to pick up the test RV with the recovery hook. Once the pick-up is successful, the test RV will be dropped and the Trieste will back off about 100 yards to check the sensitivity of the pinger on the test RV. If the sensitivity is inadequate, the Trieste can follow the "trail ball" mark on the ocean bottom and return to the test RV. The RV will again be picked up with recovery hook and brought to the surface. A complete dry run of the transfer to the White Sands is planned. After completion of the test, the White Sands (with Trieste aboard) and Apache will head for Hawaii.

Sep. 29 – "Predive preparation tests" were started early this morning. Due to a series of problems, the predive sequence was completed six hours later than planned. After releasing the Trieste tow and service lines, the dive commenced at 3:45 P.M. The Trieste descended as planned and reached the bottom (4200 feet) in 45 minutes. Upon reaching the bottom, the pilots attempted to cut the test RV loose but experienced much difficulty due to lack of tension in the line. They finally succeeded and moved away from the RV to test their ability to locate it with the pinger. They were able to return to the ship and proceeded to position the recovery hook over the test RV to pick it up. This proved to be extremely difficult because of the lack of depth perception out of the view port. Although the Trieste crew was able to come close to the test RV, they did not succeed in lowering the hook over the test RV. During these maneuvers, the winch cable jumped off the pulley and in the next attempt to operate the winch the cable parted dropping the recovery hook to the bottom. Having lost 900 pounds of weight, the Trieste immediately ascended toward the surface. The fathometer indicated that the Trieste went up 400 feet before sufficient gasoline could be released to stop the ascent. When the Trieste returned to the bottom, the hook and the test RV were not in sight. A search pattern was initiated and after 45 minutes, they were located. It was then decided that the Trieste crew should attempt to pick up the hook with the mechanical manipulator jaw and surface, with the hook hanging straight down below the Trieste. This was successfully accomplished and the hook brought to the surface. The Trieste crew finally boarded the White Sands about 2:00 A.M. the following morning.

Sep. 30 - A conference was held in the morning to determine the course of action for the remainder of the test. The day is being spent repairing the hook and the Trieste equipment. The hook is being repainted with white stripes to improve the visibility underwater. The Trieste is being kept in tow.

Oct. 1 - Activity on the planned dive came to a standstill because it was discovered that the White Sands fresh water supply is contaminated with sea water. The White Sands was ordered back to the Naval Base. The Trieste will be towed by the Apache.

Second Trieste Test Dive Off California:

Oct. 5 - The White Sands fresh water supply was replenished (40,000 gallons) and shot is being loaded into the Trieste (10,000 pounds). The White Sands, the Apache, and the Trieste will be underway to the test area this morning.

Oct. 6 - The ships returned to the test site and a second dive was completed by the Trieste. The crew is unable to pinpoint the test RV due to a failure of the underwater computer. The Trieste surfaced about five miles from the target location. The White Sands crew is unable to pinpoint the location of the test RV from the surface. (While this was happening, the DeSteiguer, the search ship at the Hawaiian recovery site, was also experiencing difficulties locating the site of the lost RV-3.)[14,15]

Recovery Hook Preoperational Inspection

Third Trieste Test Dive Off California:

Oct. 8 - The Trieste made a third attempt to recover the test RV. After maneuvering to the test RV, it was discovered that the battery power was running low on the Trieste. The crew exercised the hook in the normal recovery mode without actually maneuvering over the test RV. The operation was successful and the Trieste crew is now confident that the test RV can be picked up with the hook given sufficient time. After

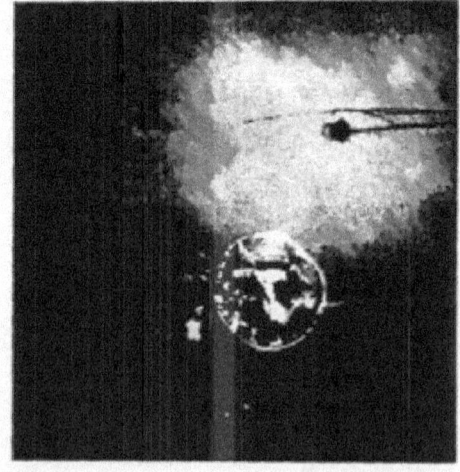

RV-3 Resting on Ocean Bottom

surfacing, the Trieste was loaded aboard the White Sands.[16]

Oct. 13 - After stowing all the equipment on the White Sands and the Trieste, the task force headed for the Hawaiian recovery area.

Oct. 20 - The search team aboard the DeSteiguer located the lost RV-3 and photographed it at a depth of 16,400 feet.

Oct. 28 - The White Sands and the Apache arrived near the recovery area and are stationed 300 miles off Hawaii.

Oct. 31 - The Maxine-D, a sea-going support ship, put out to sea at 3:00 P.M. from Pearl Harbor, Hawaii to rendezvous with the White Sands. The ship carried representatives from the customer's office, Perkin-Elmer, and five sailors who were to transfer to the White Sands crew.

Nov. 1 - The Maxine-D proceeded on schedule and reached the White Sands about 4:00 P.M. The White Sands embarked a 16-foot whaler and the representatives and sailors were transferred to the White Sands. Sea swells about eight to ten feet made the transfer extremely difficult. No casualties occurred but most of the luggage got wet. As the Maxine-D turned back to head for Hawaii, her signal flags read "THINK DEEP". The White Sands responded with "THINK DEEPER."[17]

Nov. 2 - The White Sands is approaching the recovery site and preparations for the first dive to recover the lost RV-3 are being made. The refrigerator box for cooling the recovered RV is in the final stages of completion. The Trieste was prepared for launching and about 8:30 P.M., the dock well on the White Sands was flooded. By 11:30 P.M., the Trieste was trailing in tow, off the stern of the White Sands.

The White Sands

Nov. 3 - The gassing operation, pumping 67,000 gallons of aviation fuel into the ballast tanks of the Trieste, was completed by 9:30 A.M. Soon after, 32 tons of steel shot in a slurry of sea water were pumped into the Trieste ballast tanks. Meanwhile, the RV container was lowered into the completed refrigerator box and the temperature was brought down to about 56°F. Since this was not low enough, an air drop of additional ice was requested.

First Trieste Recovery Dive Off Hawaii:

Nov. 4 - An aircraft approached the White Sands, and after dropping a smoke marker, it made six successive passes parachuting capsules containing ice. These were recovered by a small boat and brought aboard the White Sands. After the additional ice was added to the refrigerator box, the temperature stabilized just below 50°F. As the White Sands approached the RV-3 recovery site, attempts were made to locate the transponders laid by the DeSteiguer search team to mark the location of the lost RV-3. The transponder is a transmitter which returns a signal when it is interrogated by a signal on a proper frequency. The DeSteiguer team planted two such transponders (dots). Dot zero is 165 yards north of the RV-3 and dot 3 is 110 yards northeast of the dot zero. These dots were located from the White Sands using satellite navigation and then interrogated from the Apache, the sea-going support ship. Five additional dots were planted by the Apache to act as position markers. Final preparations for the first RV-3 recovery dive began as the White Sands positioned itself near the 0 Dot planted by the Apache. The 0 Dot is within 30 yards of dot zero (planted by DeSteiguer). The recovery hook was attached to the Treiste and the Trieste chambers were filled with sea water. As the Trieste descended towards the bottom, hydrophone communication was maintained throughout the dive. At about 8:00 P.M., one hour and forty-five minutes after leaving the surface, the Trieste reported their position 300 feet above the bottom which was at 16,400 feet. Relaying their range to each of the transponder dots in the pattern, the Trieste attempted to close in on dot zero. As the Trieste approached dot zero, they observed a sonar contact. The Trieste crew changed course to investigate the contact but found nothing. Other search maneuvers were conducted with several sonar contacts, but each time no visual contact was made. At about 2:00 A.M. the following morning, the dive was terminated and the Trieste reached the surface about 3:45 A.M.

Nov. 5 - A meeting was held aboard the White Sands to brief all parties concerning the observations made on the first dive. The meeting was terminated early to allow the Apache officers to return to their ship because the weather was getting worse. The plan to survey the RV-3 site was postponed because the White Sands could no longer maintain headway against the wind and the sea. Apache was required to provide a tow to maintain control over the Trieste which is being towed by the White Sands. All recovery activity ceased pending improvement in the weather.

Nov. 12 - Weather conditions prevented any recovery activity during the last seven days and the task force is now underway to Pearl Harbor. The seas are somewhat calmer but still running about four to six feet. Winds are from 10 to 15 knots.[18]

Nov. 15 - The task force was met by three tugs at the mouth of Pearl Harbor just after noon. The Apache was uncoupled and the White Sands proceeded to a pier on Ford Island. This island is in the middle of Pearl Harbor.

Nov. 16 - A briefing was held at the submarine Base and a decision was made to return to the recovery area of Hawaii on 20 November. This time, however, more support will be available from the Navy.

Nov. 19 - Provisions are being loaded and a new refrigerator was brought aboard. The unit is self-contained and capable of maintaining temperatures as low as zero degrees. The plan is to use a new towing ship, the Coucall, to tow the White Sands to the recovery site, to be joined later by the Apache.[19]

Nov. 20 - A critical problem with the number two boiler on the White Sands delayed sailing time until tomorrow.

Nov. 21 - The White Sands and the Coucall are underway. A problem

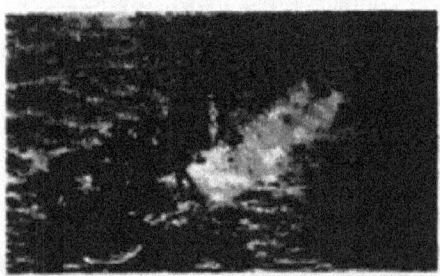

Divers Entering Water to Attach Recovery Hook to Trieste

with Coucall's towing engine resulted in the towing line separating from the White Sands. The Apache was immediately dispatched to pick up the White Sands tow. Underway to recovery site by nightfall.

- Nov. 24 – Reached the RV-3 recovery site early this morning. A search commenced for the transponders. By early evening dot 3 and dot zero were located. For the next five days the weather again prevented any recovery activity.

- Nov. 29 – Gassing of the Trieste was completed in record time but final preparations for the second dive will not be attempted until tomorrow.

Second Trieste Recovery Dive Off Hawaii:

- Nov. 30 – Steel shot was loaded aboard the Trieste this morning and electrical checkout started about 8:00 A.M. This operation, however, was slowed down by the presence of a number of uninvited guests. About a dozen large sharks and a 20-foot killer whale came over to investigate the activity around the Trieste which was now in tow. Although this did not stop the divers completely, it slowed them down considerably because while two divers worked, one kept an eye out for the sharks. All the electrical problems were corrected and the recovery hook was lowered into the water. The dive got underway at 5:45 P.M.. The Trieste continued its dive to about 15,000 feet and started interrogating the transponders. The Trieste crew pinpointed their location (5000 feet from #3 Dot). After heading for the target and lowering to the bottom, they searched the base line between dots zero and #3. At this point the computer power supply failed and dumped part of the navigation memory. The doppler sonar system also failed. However, the loss of this capability did not result in aborting the recovery attempt since the crew was able to determine their position with velocity-time relationships (dead-reckoning). After being down for almost eight hours, the Trieste crew picked up a signal in a direction west of dot #3. They headed in that direction until they suddenly lost contact. They immediately slowed down and peering out the viewing scope they spotted the lost RV-3 passing about two feet starboard. They came to an immediate stop but their momentum carried them beyond the RV-3. They then started to maneuver the Trieste around to bring RV-3 into view when the low voltage alarm sounded. They tried to plant another transponder next to the RV but were unable to do so because the mechanical arm would not operate properly at low voltage. With very little power remaining and little hope of tripping the recovery hook, the Trieste headed for the surface. They reached the surface at 4:14 A.M. the following morning. (Unfortunately, during preparations for the next dive the weather became worse and after being on station for the next seven days the task force headed for a spot in the lee of the islands between Oahu and Kauai. After some equipment repairs, the recovery team returned to the RV-3 site and remained on station.[20] High seas continued for the next few days and the recovery task force returned to Hawaii where the White Sands was placed in dry dock until 15 March 1972. After completion of repairs the recovery task force returned to the RV-3 search area on 1 April 1972.)

Documentation of the final efforts at the recovery site were not made available to Perkin-Elmer so it is not possible to record the final recovery effort in detail. However, it was learned that the Trieste did finally recover the RV-3, attach the hook, bring the RV-3 near the surface where a crane on the White Sands was to lift the RV and load it into the container. However, as the RV was lifted near the surface, it literally split apart due to internal water pressure and weight of the flooded RV. Although the hooked structure was salvaged, the film load fell back into the ocean where it presumably still lies today.

It would be interesting to determine the total cost of the nine-month recovery operation. The cost for ten days of search time by the DeSteiguer team plus four days of travel time, was estimated to be $100,000. Whatever the total cost, the recovery effort was extremely important since the RV-3 could not now be recovered by the Russians. This was of some concern because during the recovery operations, a Russian "fishing" trawler was always to be seen on the horizon by the White Sands crew.

The persistence and courage of all personnel involved in the recovery was commendable.

The Captain and crew of the Trieste II deserve a special mention for their ability and courage in continuing the search for the lost RV-3 under extremely difficult conditions.

THIS PAGE INTENTIONALLY LEFT BLANK

6 EPILOGUE

On July 12, 1983, Flight Model No. 20 was shipped to the West Coast. On that same day, a commemoration was held in the high bay area of the Wooster Heights Byeman Facility, Perkin-Elmer, Danbury, Connecticut.

Over 400 Perkin-Elmer and government people who participated on the development of the Sensor Subsystem for the Hexagon Program gathered to hear several speakers from both government and Perkin-Elmer management. The following is the speech given by Robert H. Sorensen, Chairman of the Board of the Perkin-Elmer Corporation.

"Good afternoon. This past month marked the beginning of the twentieth year of Perkin-Elmer's involvement in the overhead reconnaissance program known as Hexagon; and, we still have at least five more important years to go. That quarter of a century is a long time! At the end of that 25-year period the Hexagon Program will have spanned more than half the lifetime of the Perkin-Elmer Corporation and produced photographic imagery which would cover the world over 18 times. Successes such as the Hexagon Program have helped Perkin-Elmer expand twenty-fold in these past twenty years. In hindsight that might be described as 20/20 vision.

The brief history I am about to recount justly deserves the theme of today's meeting - "We Met the Challenge". In saying "we", I clearly intend to include our customers, co-contractors, suppliers, and the people from Perkin-Elmer who collectively have contributed to the success of the Hexagon Program.

We are pausing today to commemorate your achievements over these past twenty years and, more specifically, to acknowledge the shipment earlier this morning of the twentieth flight Hexagon Sensor Subsystem. Were this ceremony held outdoors, Hexagon System Number 18, now in orbital operation, could record this event in fine detail. Hexagon is a truly remarkable program and without question one of the foremost astronautical engineering accomplishments of all time. Its

Robert H. Sorensen, Chairman of the Board of the Perkin-Elmer Corporation in 1983

Members of the Perkin-Elmer Hexagon Program Team

value as a national asset has been demonstrated time and time again. Without Hexagon there would not be, nor could there be, any strategic arms limitation treaties or nuclear reduction discussions. Hexagon, as this country's national means of verification has not only kept America safe for democracy but has also served to lessen tensions throughout the world. It has been doing this since its first flight in 1971 and will continue to do this through its last flight in 1987. Our country's ability to conclude the Salt I Agreement was a direct result of our mission success with SV-1.

As we pause on this commemorative day, it is appropriate not only to look back upon past challenges, but also to look forward to the challenges awaiting Hexagon in the future. For the past twenty years this system has been evolving and growing and surpassing itself in each successive mission. A system designed for a forty-five day orbital life now provides 220 day coverage with potential for an even longer life. A system designed for black and white film has accommodated a myriad of spectrally sensitive film types and emulsions. A system designed to take pictures now also maps the world. But one thing hasn't changed. A program team staffed by an exceptional collection of individuals is still staffed by an exceptional collection of individuals. Your dedication, perseverance and professionalism is what makes Hexagon work. The Company knows this and the Company wants you to know that it knows this.

The Hexagon Program has truly been a team effort, and like all good teams it has had able leaders, each of whom had his share of frustration, triumph and tedium. Perkin-Elmer's involvement began in Wilton in 1964 with a study contract. These studies lasted through 1966 and culminated in our proposal submission in July 1966 for what has come to be called Hexagon. These early studies were known as the Fulcrum Program and were headed at different times by Earle Brown, Milt Rosenau, Ken Macliesh and Mike Maguire. During that time, it was my privilege as Manager of the Optical Group to accept the challenge given to Perkin-Elmer to define a space reconnaissance sensor that could meet the Hexagon requirements. On October 10th, 1966, we received word that Perkin-Elmer had been selected for the Hexagon Program. Three weeks later we announced the Company's plan to

Loading of Flight Model 20 into the Transporter

Transporter Being Loaded into a C5A Military Transport

construct this Wooster Heights Facility and the Optical Technology Division was established with Dick Werner in command to implement the Hexagon Program under the watchful eyes of the Government Program Director, Don Patterson. We were ready! The initial contract was for six units, later to be designated Block I. While Wooster Heights was being built, work on the program progressed at 77 Danbury Road in a building purchased specifically for that

C5A Military Transporter Taking Off with Flight Model 20 from Bradley International Airport

purpose. In 1967 Ken Patrick took over OTD with Mike Maguire as his Program Manager. Ken and Mike piloted the program through PDR and CDR and in February of 1968 they led the move to Danbury.

Ken was succeeded as OTD General Manager in August 1969 by Mike Maguire who, assisted by the likes of Paul Petty, Arnie Wallace, Charlie Karatzas, Bob Jones, and Harvey Henderson, guided the program through the seemingly unending frustrations of film that folded, creased, crinkled, cracked, split, broke, and, in general, refused to track.

Many of you will recall the models which were built for the abbreviated film path and the frustration we all felt during the critical periods of testing and retesting. In looking back, it is quite understandable that Dr. Al Flax, who then was in the position which Mr. Aldridge occupies today, would visit us periodically and frequently. he was quite prepared to stipulate that a fine-optical company like Perkin-Elmer could fabricate, test, and mount the optics, but could we ever devise a suitable subsystem to handle film. But in the period of patient understanding and with the collective contributions of many, many people, we met the challenge. We also were visited by the Chairman of the President's Scientific Advisory Committee, Dr. Land, who confirmed that Don Cowles' invention, the twister, would permit the handling of film off the optical bar, which proved to be an effective method of handling film in the Hexagon configuration. After convincing proof of these critical developments, we built both the engineering and development models and delivered the development model in April 1970.

In July of that year Production Unit 2 was substituted for Production Unit 1 as the lead vehicle. It was this vehicle which, after spending two months in and out of Chamber A, behind you, was shipped west on October 19th, 1970 where our West Coast Field Office was ready to perform the mate-up of the forward film path, to perform the photographic verification of flight readiness, and to support the launch and flight operations. They too met the challenge. Eight months later, at 1:41 p.m. EST on June 15th, the first vehicle was successfully launched into polar orbit from Vandenburg Air Force Base. The age of Big Bird had begun!

SV-1, as the first vehicle was designated, was kept in orbit for 31 days during which it transported 175,601 feet or over 30 miles of 6 inch wide black and white film. It photographed 32 million nautical square miles of earth; one million square miles for every day in orbit and roughly one-half of the 43 million nautical square miles of land mass on this planet. We again met the challenge.

In April of 1971, six additional units, Block II, were ordered. The Government Program Director, Don Patterson, passed the baton to Don Haas.

1972 saw the launch and successful operation of three Hexagon vehicles - in

Hx ~~TOP SECRET~~

January, July and October. The October flight, SV-4, was the first to fly color film.

1972 also marked the opening of our Los Angeles Field Office. As we got experience under our belt the flights got longer and longer - SV-4 at 69 days was more than twice as long as SV-1. the Year ended with an order for six more units, Block III.

By 1973 Mike Weeks was in charge of OTD with B. Alan Ross as Program Director. This year also saw three flights - all successful and each one longer than its predecessor. SV-7, launched in November, marked the beginning of Block II flights and achieved the first Hexagon 100-plus day orbital operation. In July of '73 customer management of the Hexagon Program was transferred from the East Coast to the Special Projects Office of the Secretary of the Air Force located in Los Angeles commanded by General Lew Allen. Colonel Ray Anderson assumed the position of Government Program Director. Thus begun a long and fruitful relationship which is ten years old this month.

General Dave Bradburn succeeded General Lou Allen.

1974 saw the launch of SV-8 and yet another milestone, the first re-flight of a recovered take-up assembly. Responsibility for refurbishing take-ups was assigned to OTD in '72 and we have been routinely recycling and reflying this recovered hardware ever since.

Seventy-four saw the incorporation of Redirection I which stretched out delivery and development of the remaining units and started the program to design both the Solid State Stellar Cameras (S^3) and the large looper. SV-9, launched in october '74, performed the first experiment using stellar photography to demonstrate the calibration technique vital for using Hexagon to map the world.

In 1975 General Jack Kulpa assumed command of the Special Projects Office. Nineteen seventy-five is remembered for the incorporation of the field contract into Block II, the appointment of Paul Petty as OTD general Manager and Bernie Malin as Program Director. In that same year, the earth was photographed by SV-9, SV-10, and SV-11. One-month missions were now four-month missions, each flight introduced a new film type, problems arose and were resolved, and the program was redirected for a second time. Over 500 people were on the program at OTD, working long hours, putting the program ahead of personal commitments, and contributing fully to the by now proven program philosophy of: doing whatever is necessary to achieve the ultimate in system performance. This philosophy resulted in an era of unprecedented program success beginning with one thousand plus photo operations of SV-12 and continuing to the present.

The launch of SV-12 in July of '76 was the only launch that year, but its duration was longer than the first three vehicles combined. The Hexagon Program was ten years old and in full bloom. S^3 and the large looper were incorporated into the sensor subsystem and Redirection III stretched out the program to one flight a year of six months' duration. These developments represented a fundamental change in both system configuration and system mission. The task of guiding Hexagon through these changes fell to Mike Mazaika and Ken Meserve as Program Directors, and Jack Rehnberg as OTD General Manager. In 1978 Colonel Les McChristian succeeded Colonel Ray Anderson. In 1978 both Redirection IV and Block IV were negotiated which further extended the program by two additional vehicles and four additional years.

While the major changes were being incorporated into SV-17, Big Birds were still flying. SV-13 in '77, 14 in '78, 15 in '79; each staying up longer, performing more operations and transporting more film. SV-15 transported 120,000 more feet of film than SV-1. That's 22 extra miles of film, enough to photograph 23 million nautical square miles.

The new decade began with the launch of SV-16 in June 1980, a mission that was to last 261 days and be distinguished by the positioning of the vehicle into a parking orbit for 90 days. SV-16 was the end of the era for the small looper; this vehicle's fantastic success heightened expectations for SV-17. It was only fitting that the chief proponent of S^3 (Solid State

Colonel Larry Cress, United States Air Force, SAFSP, Hexagon Program Director, addressing the Perkin-Elmer team in July 1983.

Stellar Camera) be put in charge of Hexagon in time for its first mapping operation, thus Vic Abraham became Program Director in 1980, and, just to show how time flies when you're having fun, Vic has held this position longer than any of his predecessors.

The launch of SV-17 in May, 1982, signalled a new age in the Hexagon chronicle - that of the Metric Pan Camera System. The large looper reduced interoperation film space, i.e., film waste, from 18% to 2.6%. With vehicle SV-17 two thousand, one hundred photo operations were made, 50 percent more than SV-16 and 5 times as much as SV-1. Furthermore, the Solid State Stellar Cameras would now permit Hexagon to map the world. We met the challenge!

In 1983 General Ralph Jacobson assumed command of the Special Projects Office and Colonel Larry Cress succeeded Colonel Les McChristian as Program Director.

Through all of these years the Hexagon team was guided by the various very able Government Program Managers Roy Burks, Bob Kohler, Colonel ▬▬▬, Colonel Dave Raspet, Colonel ▬▬▬, Major ▬▬▬ up to the present Program Manager, Major ▬▬▬.

As I said earlier, today is not only a day to look back, but also a day to look ahead to the future challenges awaiting Hexagon. The first twelve Hexagon units, i.e., all of Block I and Block II combined, flew for a total of 1065 days. The potential operating life for SV-18, 19 and 20 approaches that. This program still has a long way to go! We cannot and simply will not become complacent or disinterested! The skills and expertise developed on the Hexagon Program have been, are now, and will be, used to support this Country's needs, goals, and aspirations both in space and on the ground. We therefore intend to strive for excellence in the conduct of the Hexagon Program by keeping our proven program philosophy alive during the mission years of vehicles 18, 19, and 20.

My concluding remarks are directed mainly but not solely to the Perkin-Elmer people here today. You can be justly proud of your achievements. The challenge put forth in 1965 has been met, the National Defense has been served, the prospects for peace in the world has been enhanced - thanks in part to your continuing dedication to this vital national resource. on behalf of the Management of the Corporation, I extend our sincere thanks for a job well done and a challenge well met. Let's meet our new challenges with the same style. Thank you."

A Memento Presented to All Personnel on the Hexagon Program Team

THIS PAGE INTENTIONALLY LEFT BLANK

7 HEXAGON IMAGERY

APPENDICES

APPENDIX A
FLIGHT PERFORMANCE RECORD

Block	Flight	Launch Date	Duration (Days)	Film Transported (Feet)	Photo Operations	Average Ground Resolution Feet
I	1	6-15-71	31	175,601	430	3.5 (1414)
	2	1-20-72	41	156,192	426	2.8 (1414)
	3	7-7-72	57	185,325	720	3.1 (1414)
	4	10-10-72	69	218,346	731	3.4 (1414)
	5	3-9-73	64	218,338	639	2.4 (1414)
	6	7-13-73	74	212,432	666	2.9 (1414)
II	7	11-10-73	103	213,633	700	2.7 (1414)
	8	4-10-74	106	215,693	774	2.5 (1414)
	9	10-29-74	130	228,619	759	2.1 (1414)
	10	6-8-75	121	225,691	789	2.3 (1414)
	11	12-4-75	117	167,685	850	2.4 (1414)
	12	7-8-76	155	237,425	1,048	2.3 (1414)
III	13	6-27-77	176	239,331	1,068	2.6 (1414)
	14	3-16-78	178	249,630	1,471	2.1 (SO-208)
	15	3-16-79	188	295,750	1,501	(SO-315)
	16	6-18-80	261	295,248	1,442	(SO-315)
	17	5-11-82	204	303,527	2,112	(SO-315)
	18	6-20-83	270	307,733	1,787	2.0 (SO-315)
IV	19					
	20					

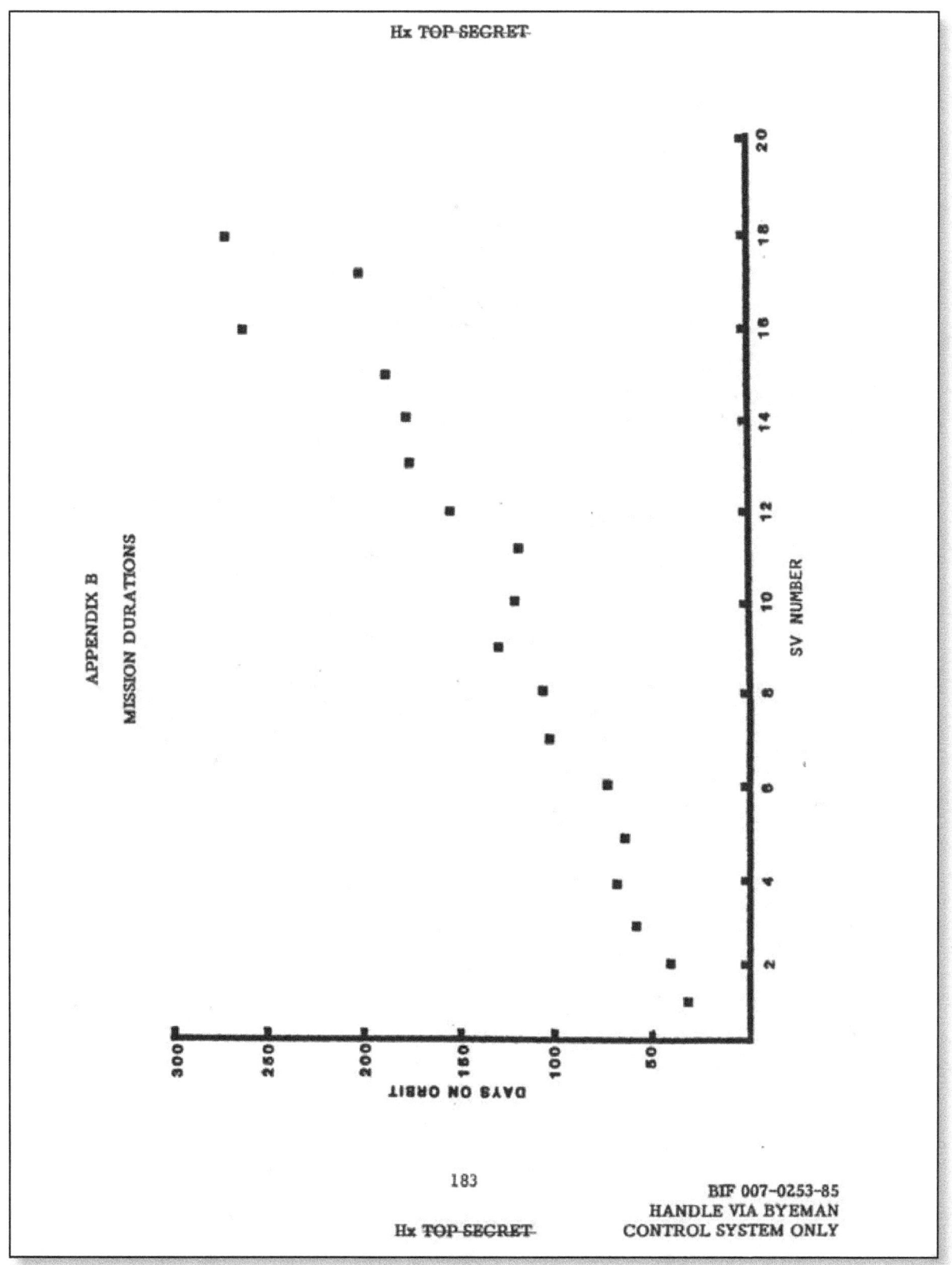

APPENDIX B
MISSION DURATIONS

APPENDIX C
MISSION OPERATIONS

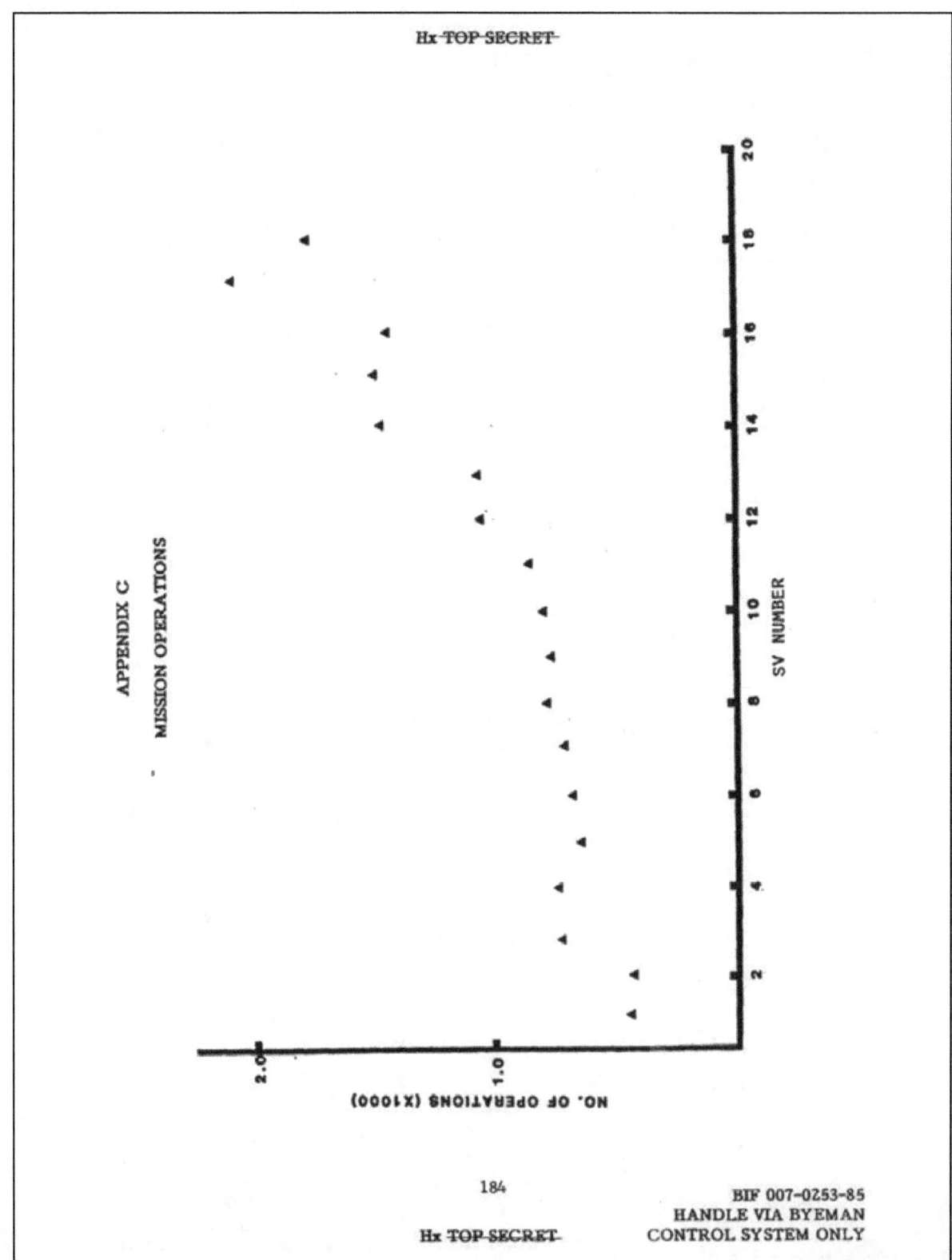

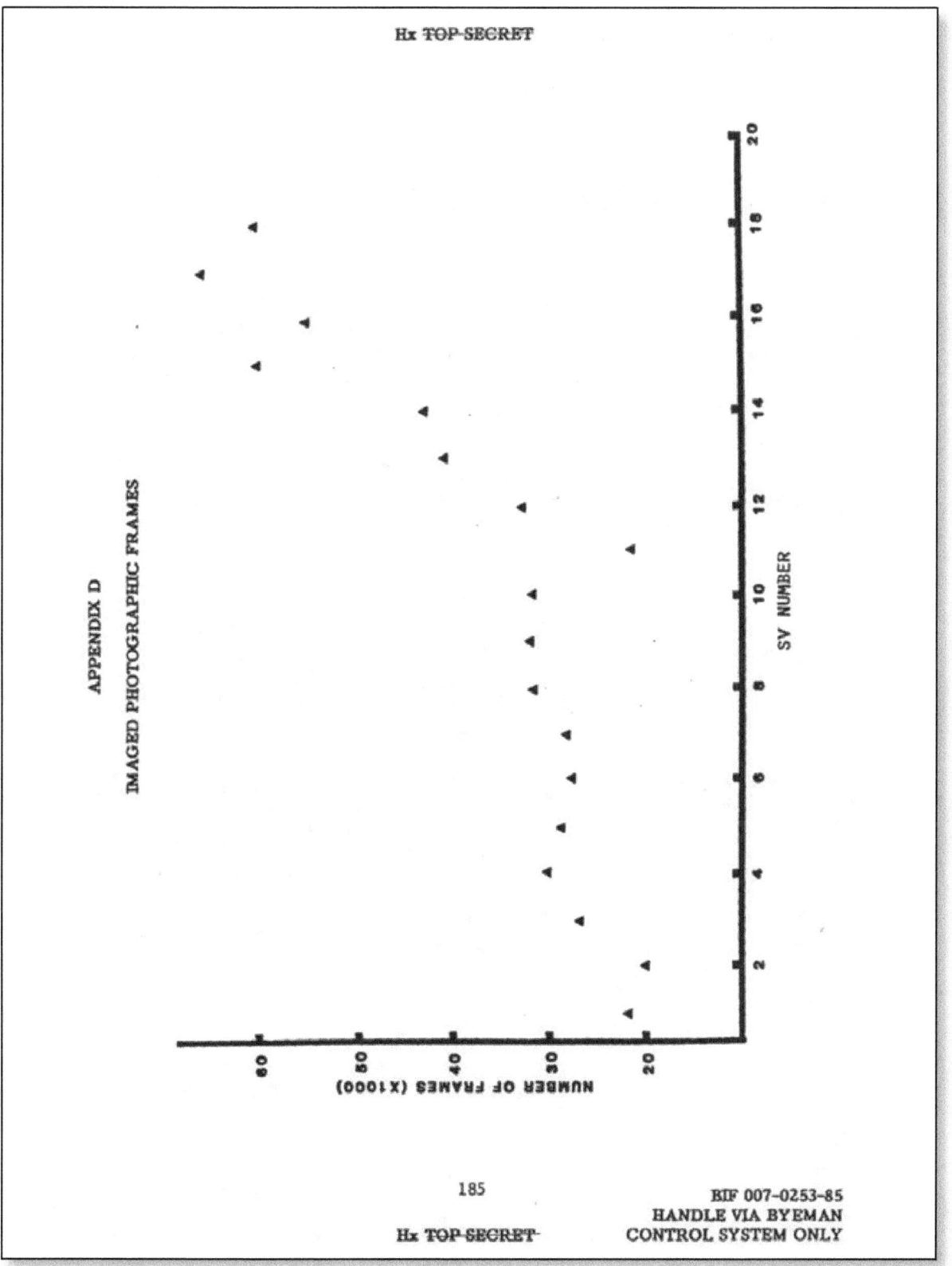

APPENDIX D

IMAGED PHOTOGRAPHIC FRAMES

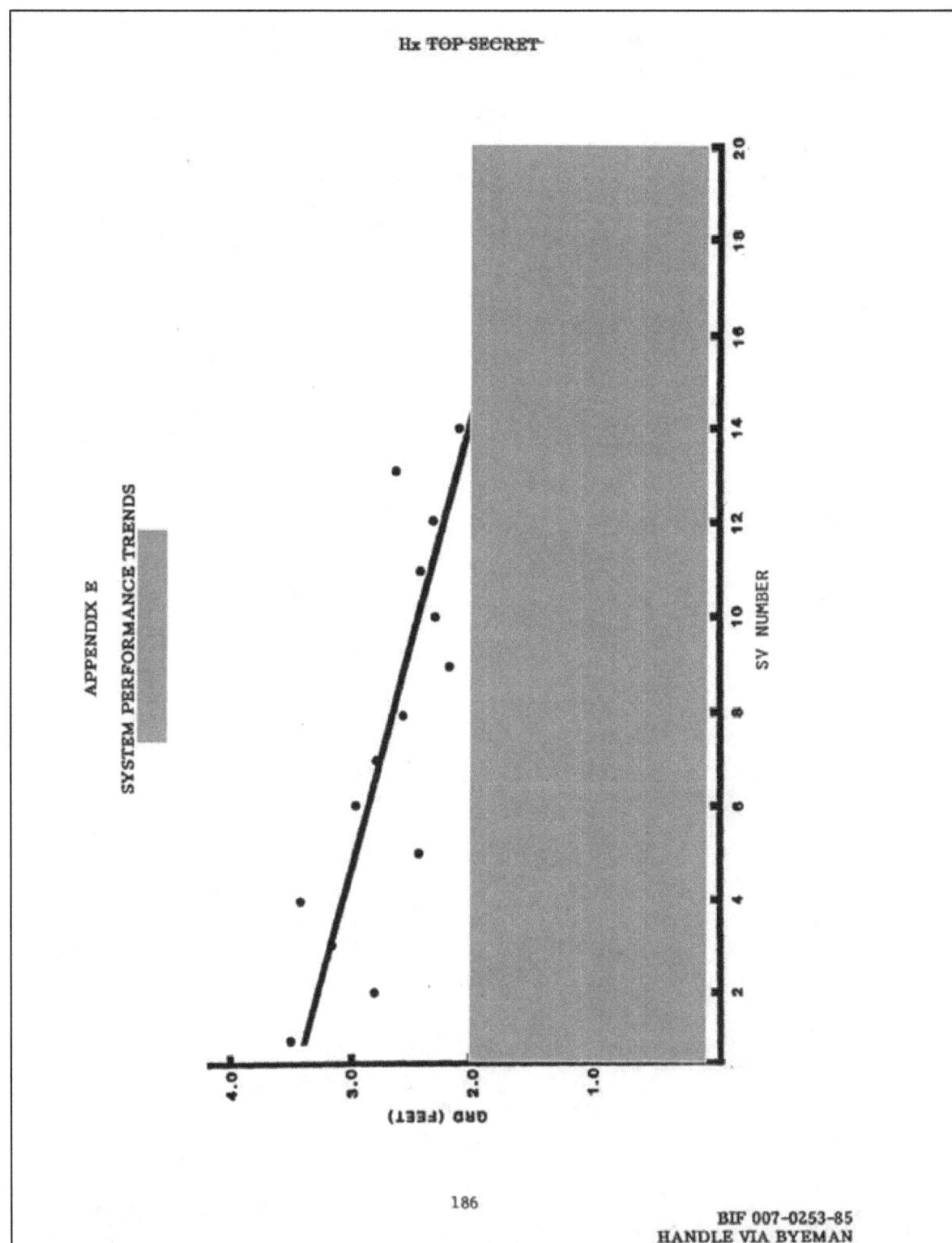

APPENDIX E
SYSTEM PERFORMANCE TRENDS

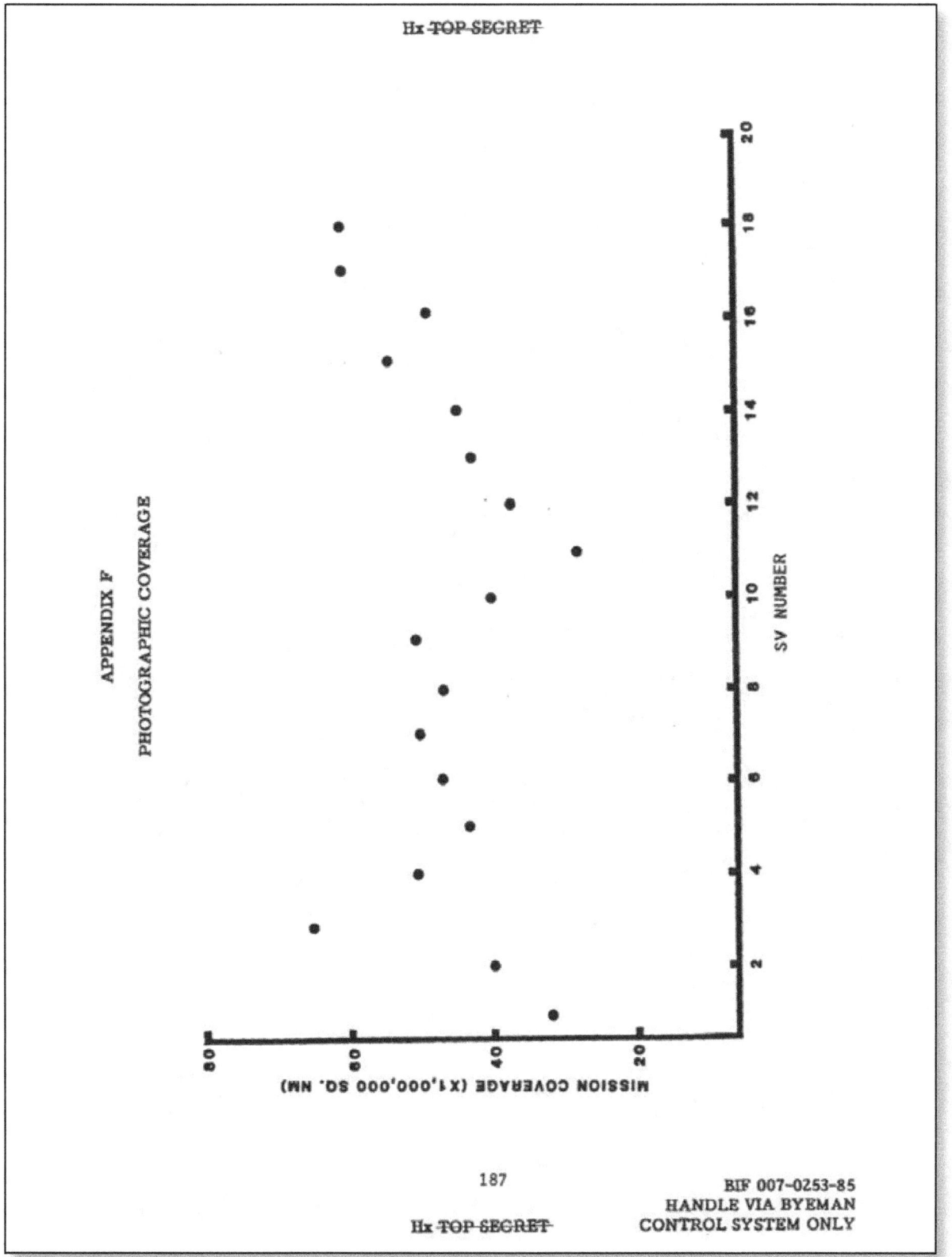

Hx ~~TOP SECRET~~

APPENDIX G
LISTING OF POST FLIGHT REPORTS

Mission Numbers	Document Numbers
1201 (S/N 003)	BIF 007-1325-71
1202 (S/N 002)	BIF 007-0460-72
1203 (S/N 004)	BIF 007-1231-72
1204 (S/N 005)	BIF 007-0158-73
1205 (S/N 006)	BIF 007-0662-73
1206 (S/N 007)	BIF 007-1247-73
1207 (S/N 010)	BIF 007-0405-74
1208 (S/N 011)	BIF 007-0855-74
1209 (S/N 012)	BIF 007-0258-75
1210 (S/N 013)	BIF 007-0711-75
1211 (S/N 014)	BIF 007-0333-76
1212 (S/N 015)	BIF 007-0105-77
1213 (S/N 016)	BIFX 007-3781-78
1214 (S/N 017)	BIFX 007-3290-79
1215 (S/N 018)	BIFX 007-3153-80
1216 (S/N 019)	BIF 007-0211-81
1217 (S/N 020)	BIF 007-0238-83
1218 (S/N 021)	BIF 007-3231-84
1219 (S/N 022)	BIF 007-3231-84
1220 (S/N 023)	

Hx ~~TOP SECRET~~

BIF 007-0253-85
HANDLE VIA BYEMAN
CONTROL SYSTEM ONLY

Hx TOP SECRET

APPENDIX H

ORGANIZATION CHRONOLOGY

A PARTIAL LIST OF PERKIN-ELMER OFFICERS AND MANAGERS MENTIONED IN THIS DOCUMENT

Chairman of the Board

Richard S. Perkin
August 25, 1958 – May 22, 1969

Chester W. Nimitz, Jr.
July 31, 1969 – February 28, 1980

Robert H. Sorensen
March 1, 1980 – March 21, 1985

Horace G. McDonell, Jr.
March 21, 1985 –

Presidents

Richard S. Perkin
December 13, 1939 – June 1, 1961

Robert E. Lewis
June 1, 1961 – December 31, 1964

Chester W. Nimitz, Jr.
January 1, 1965 – February 15, 1973

Robert H. Sorensen
February 15, 1973 – March 1, 1980

Horace G. McDonell, Jr.
March 1, 1980 – March 21, 1985

Gaynor N. Kelley
March 21, 1985 –

General Managers, Electro-Optical Division

Dr. Roderic M. Scott
August 1, 1956 – July 31, 1958

Carlton W. Miller
August 1, 1958 – July 31, 1960

Robert H. Sorensen
September 19, 1960 – July 31, 1965

Kennett W. Patrick
August 1, 1965 – July 31, 1967*

Note: EOD was formed in 1956 when the company divided into the Instrument Division and the Engineering and Optical Division (later changed to Electro-Optical Division) and included the Engineering Branch in which the Fulcrum program was initiated.

*EOD listing to this date only.

Hx TOP-SECRET

APPENDIX H (Continued)

General Managers, Electro-Optical Group

Robert H. Sorensen, Vice President
June 10, 1965 - July 31, 1966

Note: The Electro-Optical Group was formed in 1965 and included the Electro-Optical Division and the Astro-Optical Division. The name of this group changed to the Optical Group when the Optical Technology Division was formed January 28, 1966.

General Managers, Optical Group

Robert H. Sorensen, Senior Vice President
May 27, 1966 - February 15, 1973

Dr. Donald A. Dooley, Sr. Vice President
May 6, 1974 - November 18, 1976

Edward F. Ronan, Sr. Vice President
August 31, 1976 - March 1, 1983

Gaynor N. Kelley, (Acting) Executive Vice President
March 2, 1983 - June 21, 1983

William W. Chorske, Sr. Vice President
June 22, 1983 - April 24, 1985

Mercade A. Cramer, Sr. Vice President
April 24, 1985 -

General Managers, Optical Technology Division

W. Richard Werner
October 10, 1966 - December 31, 1966

Kennett W. Patrick, Vice President
January 1, 1967 - August 23, 1969

Michael F. Maguire, Vice President
August 24, 1969 - January 14, 1973

L. Michael Weeks, Vice President
January 15, 1973 - July 12, 1975

Paul E. Petty, Vice President
July 13, 1975 - July 31, 1978

John D. Rehnberg, Vice President
August 1, 1978 - July 31, 1981

Paul E. Petty, Vice President
August 1, 1981 -

BIF 007-0253-85
HANDLE VIA BYEMAN
CONTROL SYSTEM ONLY

Hx TOP-SECRET

Hx TOP SECRET

APPENDIX H (Continued)

Program Managers, Fulcrum Program

Earle Brown
June 28, 1964 – September 30, 1964

Milton D. Rosenau
October 1, 1964 – January 31, 1965

Dr. Kenneth G. Macleish
February 1, 1965 – December 8, 1965

Michael F. Maguire
December 8, 1965 – October 9, 1966

Program Managers, Hexagon Program

Michael F. Maguire
October 10, 1966 – August 23, 1969

Bernard Malin
August 24, 1969 – January 27, 1971

Paul E. Petty
January 28, 1971 – April 27, 1973

B. Alan Ross
April 28, 1973 – September 7, 1975

Bernard Malin
September 8, 1975 – December 16, 1977

Michael A. Mazaika
December 19, 1977 – July 11, 1979

Kent H. Meserve
July 12, 1979 – October 2, 1980

Victor Abraham
October 3, 1980 – January 24, 1985

Leonard J. Farkas
January 25, 1985 –

BIF 007-0253-85
HANDLE VIA BYEMAN
CONTROL SYSTEM ONLY

Hx TOP SECRET

APPENDIX I

DESCRIPTION OF
PROGRAM DOCUMENTATION

The documentation requirements of the Hexagon program changed constantly to suit the needs of both the customer and Perkin-Elmer. The initial parametric study that started in June 1964, and later the Phase 1 Fulcrum effort that began in September 1964 and ended in January 1965, needed only interim reports and occasional memos and TWX messages to report the technical progress on the program.

After Perkin-Elmer was asked by the CIA to continue the design effort started by Itek, a more formalized type of reporting was required. Starting on 27 August 1965, Perkin-Elmer began sending bi-weekly TWX progress reports to the customer. These TWX messages continued in an unbroken chain until 12 June 1968.

Technical review meetings, which started in late 1965, continued throughout the life of the program, and minutes on most of these meetings are in the archives. The Sensor Subsystem Monthly Technical Report was initiated in December 1966 (DMR-1) and is still being published today.

The Data Management List, which recorded all incoming and outgoing documentation on the program beginning in November 1966, was eventually discontinued during the first program personnel layoff in September 1970. A listing similar to the Data Management List was included in the Sensor Subsystem Monthly Technical Report.

Highlight reports began in August 1968 and continued in various forms (both TWX messages and memorandum) until September 1970. A monthly schedule status report was started on 15 December 1967, later changed into a notebook form called the "blue book".

The Hexagon contract, of course, contains a Contract Data Requirements List (CDRL) which lists program documentation requirements.

APPENDIX J
SOURCE NOTES

Foreword

1. Hexagon Mission History Summary (SV-1 through SV-17), revised April 1983, (BIF 007-0881-74I).
2. Edward C. Aldridge Speech, 12 July 1983.

Introduction

1. Harry Howe Ransom, The Intelligence Establishment (Harvard University Press, 1970), p. 48.
2. Ibid., p. 121.
3. Ibid., p. 123.
4. Philip J. Klass, Secret Sentries In Space (New York: Random House, 1971), p. 72.
5. Ibid., p. 85.
6. Ibid., pp. 91-92.
7. George L. Christian, "RB-58 Photo System Uses TV Viewer," Aviation Week, 6 January 1958, pp. 90-93.
8. TWX Message [redacted], 12 October 1966.

SECTION 1 PROGRAM OVERVIEW

Early Background

1. Richard C. Babish interview, 17 September 1980.
2. David Wise and Thomas B. Ross, The U-2 Affair (city, pub, yr), p. 48.
3. Philip J. Klass, Secret Sentries In Space (New York: Random House, 1971), p. 82.
4. Final Itek Summary Report (R9204-13), 26 February 1965 (Perkin-Elmer AH65-0769).
5. TWX Message [redacted], 7 October 1964.
6. Final Itek Summary Report (R9204-13).
7. Aviation Week, 6 January 1964, p. 87-88.
8. Ibid, p. 88.
9. Ibid, p. 85.
10. Sensor Subsystem Negotiation Handbook, 6 March 1968, P. I-4 (Perkin-Elmer DAN-68-0315).
11. Memorandum, Dr. Kenneth G. Macleish to Richard S. Perkin, 19 September 1964.
12. Richard C. Babish interview, 17 September 1980.
13. Interim Report, Ad Hoc Study, Perkin-Elmer Engineering Report No. 7818, 25 September 1964, p. 1 (Perkin-Elmer AH65-0007).
14. Ibid, p. 4.
15. Macleish memo, p. 1.

Early Background (Continued)

16. ABF-430, Second Progress Report, 16 November 1964 (Perkin-Elmer AH65-0609).
17. Macleish memo, p. 4.
18. Memorandum, Leslie C. Dirks to Earle B. Brown, 8 July 1964 (Perkin-Elmer DANX 67-6167).
19. Ibid, Attachment No. 1, p. 2.
20. Negotiation Handbook, p. I-4.
21. TWX Message ▓▓▓▓, 28 September 1964.
22. TWX Message ▓▓▓▓, 29 September 1964.
23. TWX Message ▓▓▓▓, 7 October 1964.
24. AD HOC Study Report Summary, 27 January 1965, p. 17 (Perkin-Elmer AH65-0501).
25. TWX Message ▓▓▓▓, 9 December 1964.
26. TWX Message ▓▓▓▓, 18 January 1965.
27. TWX Message ▓▓▓▓, 19 January 1965.
28. W. R. Werner interview 29 July 1980.
29. TWX Message ▓▓▓▓, 28 January 1965.
30. TWX Message ▓▓▓▓, 10 March 1965.
31. TWX Message ▓▓▓▓, 24 March 1965.
32. Lawrence E. Emmons interview, December 1981.
33. TWX Message ▓▓▓▓, 5 June 1965
34. Briefing At Associate Contractors, 1 April 1965 (Perkin-Elmer AH65-0664).
35. Memorandum, C.S. Morser to F. J. Madden (Perkin-Elmer AH65-0863, last frame, Microfiche No. 1).
36. Conversation, Maurice G. Burnett and Richard J. Chester, 25 June 1981.
37. TWX Message ▓▓▓▓, 13 April 2965.
38. TWX Message ▓▓▓▓, 5 May 1965.
39. ▓▓▓▓ Proposal, 14 May 1965 (Perkin-Elmer AH65-0817).
40. Robert R. Batchelder, Engineering Notebook (Itek), 21 May 1964 (Perkin-Elmer AH65-0619).
41. Robert M. Landsman, Technical Report No. 65-7, Comparison of Single Pass and Multi-pass Film Transports For Use With The Optical Bar, 31 March 1965 (Perkin-Elmer AH 65-0654).
42. Milton D. Rosenau, Technical Report No. 65-26, System Configuration Selection, 21 April 1965 (Perkin-Elmer AH67-1519).
43. TWX Message ▓▓▓▓, 6 July 1965.
44. TWX Message ▓▓▓▓, 31 July 1965.
45. TWX Message ▓▓▓▓, 4 August 1965.
46. Summary Report, F-Prime vs. M-Prime Recommendation, 30 September 1965 (Perkin-Elmer AH65-1086).
47. TWX Message ▓▓▓▓, 17 September 1965.
48. Macleish Organization Memorandum, 8 December 1965.
49. TWX Message ▓▓▓▓, 27 August 1965.
50. TWX Message ▓▓▓▓, 24 September 1965.
51. Payload Contractor's Presentation, 9 December 1965 (Perkin-Elmer AH65-1275).
52. Ibid, p. 1.
53. Memorandum, R. G. Clark to W. R. Werner, Progress Report, Ad Hoc Program, 16 February 1966 (Perkin-Elmer AH66-1278).

Hx TOP SECRET

Organizational Period

1. Functional Organization Chart, Corporate Level, 17 April 1964.
2. Perkin-Elmer Annual Report 1965, p. 12.
3. Electro-Optical Division, Engineering Organization Chart, 6 June 1964.
4. Special Projects Organization Chart, 11 June 1963.
5. Ad Hoc Study Report, Volume II, Management, 25 January 1965, Organization chart, p. 79 (Perkin-Elmer AH65-0506).
6. Sensor Subsystem Negotiation Handbook, 6 March 1968, p. I-4 (Perkin-Elmer DAN-68-0315).
7. Memorandum, Dr. Kenneth G. Macleish to All POD Supervisors, 8 December 1965.
8. Payload Contractor's Presentation, 9 December 1965 (Perkin-Elmer AH65-1275).
9. Sensor Subsystem Proposal, 21 July 1966, Program Plan, Volume II, TR-66-300 (Perkin-Elmer AH66-1402).

Early Technical Development

1. Robert R. Batchelder, Engineering Notebook (Itek), 21 May 1964 (Perkin-Elmer AH65-0619).
2. Supplementary Report 101, Choice of System Configuration, 20 Janury 1965 (Perkin-Elmer AH65-0522).
3. Ad Hoc Study Report Volume I, Part I, 26 January 1965 (Perkin-Elmer AH65-0502).
4. Ad Hoc Study Reports, January 1965 (Perkin-Elmer AH65-0501 to 0511).
5. Itek Final Summary Report, R9204-13, 26 February 1965 (Perkin-Elmer AH65-0769).
6. Itek Final Brassboard Status Report, 9204-TM232 (Perkin-Elmer AH65-0034).
7. TWX Message _____, 14 April 1965.
8. Perkin-Elmer Summary Report, F-Prime Recommendation, 30 September 1965 (Perkin-Elmer AH65-1086).
9. TWX Message _____, 7 December 1965.
10. Request For Proposal, Sensor Subsystem For General Search and Surveillance System, 19 May 1966 (Perkin-Elmer AH66-1400).
11. Perkin-Elmer Design Definition Sensor Subsystem, 21 July 1966 (Perkin-Elmer AH66-1401).
12. TWX Message _____, 28 July 1966.
13. TWX Message _____, 28 July 1966.
14. TWX Message _____, 2 September 1966.

Award of Contract

1. TWX Message _____, 12 October 1966.
2. TWX Message _____, 18 October 1966.
3. TWX Message _____, 31 October 1966.
4. Program Plan, Sensor Subsystem, TR-66-300-2, Volume II. Section II, 21 July 1966.
5. Interim Report, Ad Hoc Study, Perkin-Elmer Engineering Report No. 7818, 25 September 1964, p. 1 (Perkin-Elmer AH65-0007).
6. Interim Report, p. 135.
7. TWX Message _____ 29 September 1964.
8. TWX Message _____ 6 October 1964.
9. TWX Message _____ 28 January 1965.
10. Sensor Subsystem Negotiation Handbook, 6 March 1968, p. I-10 (Perkin-Elmer DAN-68-0315).
11. TWX Message _____, 24 March 1965.

Hx TOP SECRET

Hx ~~TOP SECRET~~

Award of Contract (Continued)

12. TWX Message ▓▓▓▓▓▓▓▓▓▓, 29 April 1965.
13. TWX Message ▓▓▓▓▓▓▓▓▓▓, 11 May 1965.
14. Kenneth G. Macleish interview, 9 December 1981.
15. Negotiation Handbook, p. I-14.
16. TWX Message ▓▓▓▓▓▓▓▓▓▓, 11 June 1965.
17. Perkin-Elmer July Work Statement (Perkin-Elmer AH65-0929).
18. TWX Message ▓▓▓▓▓▓▓▓▓▓, 31 July 1965.
19. Perkin-Elmer Three-Months Work Statement (Perkin-Elmer AH65-1023).
20. TWX Message ▓▓▓▓▓▓▓▓▓▓, 24 September 1965.
21. TWX Message ▓▓▓▓▓▓▓▓▓▓, 7 December 1965.
22. Negotiation Handbook, p. I-11.
23. TWX Message ▓▓▓▓▓▓▓▓▓▓, 13 January 1966.

Cover and Security Considerations

1. Ad Hoc Study Report, Management, Volume II, Section X, 25 January 1965 (Perkin-Elmer AH65-0506).
2. Request for Proposal, Sensor Subsystem for General Search and Surveillance System, 19 May 1966, pp. 002, 107 and 123 (Perkin-Elmer AH66-1400).
3. General Security Bill, 10 January 1967 (Perkin-Elmer DAN-67-0235).
4. Guard Orders, 10 January 1967 (Perkin-Elmer DAN-67-0236).
5. Classification of Project Documents, 10 January 1967 (Perkin-Elmer DAN-67-0229).
6. Security Classification Guide, undated (Perkin-Elmer DAN-67-0174).
7. Courier Proposal, undated (BIF 007-0560-73).
8. Robert A. Markin interview, 23 September 1981.
9. Payload Contractor's Presentation, 9 December 1981.

Building Program

1. Ad Hoc Study Report, Summary (Perkin-Elmer AH65-0501).
2. Ad Hoc Study Report, Management, 25 January 1965, Volume II (Perkin-Elmer AH65-0506).
3. Lawrence E. Emmons interview, December 1981.
4. Program Plan, Sensor Subsystem, Volume II (Perkin-Elmer AH66-1402).
5. Ibid.
6. Ibid.
7. TWX Message ▓▓▓▓▓▓▓▓▓▓, 29 July 1966.
8. TWX Message ▓▓▓▓▓, 11 August 1966.
9. TWX Message ▓▓▓▓▓▓▓▓, 22 September 1966.
10. Sensor Subsystem Monthly Technical Report No. 6, June 1967 (Perkin-Elmer DAN-67-1233).
11. Sensor Subsystem Monthly Report No. 7, July 1967 (Perkin-Elmer DAN-67-1447).
12. Sensor Subsystem Monthly Report No. 14, March 1968 (Perkin-Elmer DAN-68-0436).
13. Information Guide, Wooster Heights Plant, February 1968.
14. Kenneth W. Patrick memo to project personnel, 7 February 1968.
15. Robert A. Kelley interview, 22 July 1981.

Hx TOP SECRET

Sensor Subsystem Description

1. Technical Data Book Sensor Subsystem for Hexagon Program Satellite Vehicles SV-17 through SV-20, three volumes (BIF 007-0104-82, BIF 007-0105-82 and BIF 007-0106-82).
2. Hexagon Mission History Summary (SV-1 through SV-17) revised April 1983 (BIF 007-0881-74I).

First Flight of the Big Bird

1. Charles O. Bryant, Jr. interview, 20 May 1982.
2. Launch Certification for Sensor Subsystem S/N 003 SV-1, 14 June 1971 (BIFX 007-5401-71).
3. Sensor Subsystem Post Flight Report SV-1 (S/N 003) 20 August 1971 (BIF 007-1325-71).

SECTION 2   CUSTOMER RELATIONSHIPS AND INTERACTIONS

Program Management

1. Robert A. Markin interview, 23 September 1981.
2. Letter, Donald W. Patterson to C.W. Besserer, 6 October 1967 (Perkin-Elmer DANX-67-6917).
3. Letter, John J. Crowley to Chester W. Nimitz, Jr., 10 November 1967 (Perkin-Elmer DAN-67-7076).
4. Letter, Chester W. Nimitz, Jr., to John J. Crowley, 30 November 1967 (Perkin-Elmer DAN-67-1963).
5. Letter, Donald W. Patterson to Michael F. Maguire, 4 November 1968 (Perkin-Elmer DANX-68-9068).

Program Security

1. TWX Message ███████████, 1 February 1967.
2. TWX Message ███████████, 24 October 1966.
3. Letter, Brigadier General John E. Kulpa, Jr. to Paul E. Petty, 8 August 1975 (BIFX 007-4362-75).
4. Aviation Week, 23 June 1958, p. 18.
5. Aviation Week, 24 November 1958, p. 33.
6. Aviation Week, 8 December 1958, p. 31.
7. Aviation Week, 9 March 1959, p. 323.
8. Aviation Week, 20 April 1959, p. 26.
9. Aviation Week, 25 May 1959, p. 26.
10. Aviation Week, 24 August 1959.
11. Aviation Week, 16 November 1959, p. 33.

Hx TOP SECRET

Program Security (Continued)

12. Aviation Week, 10 April 1961, p. 30.
13. Aviation Week, 10 February 1964, p. 53.
14. Aviation Week, 6 March 1967, p. 116.
15. Aviation Week, 21 June 1971, p. 15.
16. Aviation Week, 30 August 1971, p. 12.
17. Aviation Week, 25 September 1972, p. 17.
18. Aviation Week, 7 July 1975, p. 21.
19. Aviation Week, 6 February 1978, p. 187.
20. Aviation Week, 10 December 1979, p. 66.
21. Aviation Week, 29 September 1980, p. 27.
22. Aviation Week, 6 October 1980, p. 18.
23. Aviation Week, 9 March 1981, p. 16.
24. Aviation Week, 1 February 1982, p. 13.
25. Amrom H. Katz, Astronautics, April, June, July, August, September, October, 1960.
26. Business Week, 4 June 1960, p. 30.
27. Amrom H. Katz, Selected Readings in Aerial Reconnaissance (Rand Corporation, August 1963) p. 2762.
28. Journal of the SMPTE, Volume 73, p. 858.
29. Business Week, 13 November 1965, p. 70.
30. Electronic News, 13 March 1967, p. 39.
31. Niel Jensen, Optical and Aerial Photographic Reconnaissance, 1968.
32. Electronic News, 2 September 1968, p. 64.
33. Electronic News, 12 August 1968.
34. Industrial Research, October 1968, p. 28.
35. Electronic News, 16 June 1969, p. 14.
36. Ted Greenwood, Reconnaissance, Surveillance and Arms Control (International Institute for Strategic Studies, June 1972), Adelphi Papers No. 88.
37. Business Week, 3 June 1972.
38. Scientific American, Volume 228, No. 2, February 1973.
39. Aircraft Engineering, February 1975, p. 15.
40. Electronic and Power, August 1978, p. 573.
41. New York Times, 1 March 1981, p. 12.
42. New Scientist, 1 October 1981, p. 36.
43. Time, 27 April 1981, p. 20.
44. Philip J. Klass, Secret Sentries in Space (New York, Random House, 1971), p. 82.
45. Letter, Brigadier General David D. Bradburn to Robert H. Sorensen, 23 May 1973 (BYE 96402-73).
46. Letter, Brigadier General Bradburn to R. Sorensen, 16 July 1973 (BIFX 007-4655-73).
47. Letter, Brigadier General Bradburn to R. Sorensen, 17 July 1973 (BIFX 007-4678-73).
48. Letter, R. Sorensen to Brigadier General Bradburn, 26 July 1973 (BIF 007-0841-73).
49. A Security Proposal for the Integration of the OTD Facility at Danbury, CT (BIF 007-0549-74).
50. Byeman Controlled Facility for the Space Telescope Program, PM-1596-X-A, February 1977 (BIF 007-0325-76-A).

SECTION 3 TECHNICAL DESIGN, MANUFACTURE AND TEST

Evolution of the Sensor Subsystem Design

Optical Bar Assembly

1. Sensor Subsystem Monthly Technical Report No. 1, December 1966, p. 2-1 (Perkin-Elmer DAN-67-0160).
2. Optical Bar Concept Review Meeting Minutes, 14 February 1967 (Perkin-Elmer DAN-67-0466).
3. Optical Bar Preliminary Design Review Meeting (PDR) Minutes, 6 June 1967 (Perkin-Elmer DAN-67-1220).
4. Sensor Subsystem Monthly Report No. 9, September 1967, p. 4-32 (Perkin-Elmer DAN-67-1691).
5. Optical Bar Critical Design Review (CDR) Meeting Minutes, 31 July 1968 (Perkin-Elmer DAN-68-1056).

Camera Support Frame Assembly

1. Frame Concept Review Meeting Minutes, 24 February 1967 (Perkin-Elmer DAN-67-0559).
2. Frame PDR Meeting Minutes, 23 May 1967 (Perkin-Elmer DAN-67-1111).
3. TWX Message ███████, 23 May 1967.
4. Project Memorandum, Graham F. Wallace, 25 May 1967 (Perkin-Elmer DAN-67-0982).
5. Memorandum, Frame Approval, Donald W. Patterson, 23 May 1967 (Perkin-Elmer DAN-67-0971).
6. Sensor Subsystem Monthly Technical Report No. 11, November 1967, p. 4-40 (Perkin-Elmer DAN-67-2022).
7. Frame CDR Meeting Minutes, 24 September 1968 (Perkin-Elmer DAN-68-1216).
8. Sensor Subsystem Monthly Technical Report No. 13, January 1968, p. 3-5 (Perkin-Elmer DAN-68-0213).

Film Drive Assembly

1. United States Patent No. 3,434,639, Transports for Elongated Material, Figures Nos. 5 and 6, 25 March 1969.
2. Charles D. Cowles interview, 29 September 1981.
3. Perkin-Elmer Drawing Number 606-10009, 24 October 1964.
4. Ad Hoc Technical Report Volume I, Breadboard Report No. 10, 180 Degree Twister Breadboard, 22 January 1965 (Perkin-Elmer AH 65-0505).
5. Supplementary Report 101, Choice of System Configuration, pgs. 21 and 22, 20 January 1965 (Perkin-Elmer AH 65-0522).
6. Twister Breadboard Test Results, Technical Report No. 65-128, 20 July 1965 (Perkin-Elmer DAN-67-0413).
7. Film Drive Concept Review Meeting Minutes, 2 March 1965 (Perkin-Elmer DAN-67-0782).
8. Film Drive PDR Meeting Minutes, 14 November 1967 (Perkin-Elmer DAN-67-2014).
9. Film Drive CDR Meeting Minutes, 28 August 1968 (Perkin-Elmer DAN-67-1308).

Hx TOP SECRET

Platen Assembly

1. Sensor Subsystem Monthly Technical Report No. 1, December 1966, p. 3-20 (Perkin-Elmer DAN-67-0160).
2. Platen Bearing Arrangement Concept Review Meeting Minutes, 6 January 1967 (Perkin-Elmer DAN-67-0230).
3. Sensor Subsystem Monthly Technical Report No. 2, January 1967, p. 3-9 (Perkin-Elmer DAN-67-0372).
4. Platen Design Concept (Preliminary) Meeting Minutes, 13 January 1967 (Perkin-Elmer DAN-67-0299).
5. Sensor Subsystem Monthly Technical Report No. 4, April 1967, p. 3-10 (Perkin-Elmer DAN-67-0853).
6. Platen Concept Review Meeting Minutes, 5 May 1967 (Perkin-Elmer DAN-67-1110).
7. Project Memorandum No. 324, Platen Concept Review, 1 May 1967 (Perkin-Elmer DAN-67-0830).
8. Sensor Subsystem Monthly Technical Report No. 6, June 1967, p. 4-9 (Perkin-Elmer DAN-67-1233).
9. Sensor Subsystem Monthly Technical Report No. 9, September 1967, p. 4-17 (Perkin-Elmer DAN-67-1691).
10. Platen Assembly PDR Meeting Minutes, 7 February 1968 (Perkin-Elmer DAN-68-0376).
11. Platen Assembly CDR Meeting Minutes, 25 September 1968 (Perkin-Elmer DAN-68-1254).

Supply Assembly

1. Sensor Subsystem Monthly Technical Report No. 1, December 1966, p. 3-38 (Perkin-Elmer DAN-67-0160).
2. Project Memorandum No. 247, Supply Assembly Weight Increase, 27 February 1967 (Perkin-Elmer DAN-67-0541).
3. Project Memorandum No. 162, Design Impact of New Specification, 30 November 1966 (Perkin-Elmer DAN-67-0118).
4. Sensor Subsystem Monthly Technical Report No. 2, January 1967, p. 3-6 (Perkin-Elmer DAN-67-0372).
5. Sensor Subsystem Monthly Technical Report No. 4, April 1967, p. 3-37 (Perkin-Elmer DAN-67-0853).
6. Sensor Subsystem Monthly Technical Report No. 5, May 1967, p. 4-24 (Perkin-Elmer DAN-67-1047).
7. Sensor Subsystem Monthly Technical Report No. 8, August 1968, pg. 4-43 (Perkin-Elmer DAN-67-1577).
8. Project Memorandum No. 535-X, Program and System Impact of a Design Change of Film Supply Reel Orientation, 24 August 1967 (Perkin-Elmer DAN-67-1518).
9. Supply Assembly PDR Meeting Minutes, 21 February 1968 (Perkin-Elmer DAN-68-0511).
10. Sensor Subsystem Monthly Technical Report No. 18, July 1968, p. 4-64 (Perkin-Elmer DAN-68-0895).
11. Supply Assembly CDR Meeting Minutes, 25 November 1968 (Perkin-Elmer DAN-68-1643).

Looper Assembly

1. Looper Concept Review Meeting Minutes, 23 February 1967 (Perkin-Elmer DAN-67-0778).
2. Fred Klein Memorandum, Looper Concept Review, 28 April 1967 (Perkin-Elmer DAN-67-0883).
3. Sensor Subsystem Monthly Technical Review No. 5, May 1967, pg. 4-8 (Perkin-Elmer DAN-67-1047).
4. C. W. Besserer Letter, Comments on Looper PDR Package, 31 July 1967 (Perkin-Elmer DANX-67-6779).
5. C. W. Besserer Letter, Steps to Improve Looper PDR Package, 26 September 1967 (Perkin-Elmer DANX-67-0813).
6. Looper PDR Meeting Minutes, 27 September 1967 (Perkin-Elmer DAN-67-1828).
7. Donald W. Patterson Memorandum, Looper PDR, 29 September 1967.
8. K. W. Patrick Memorandum, Looper PDR, 15 November 1967 (Perkin-Elmer DAN-67-1908).
9. Donald W. Patterson Memorandum, Looper PDR, 28 November 1967 (Perkin-Elmer DANX-67-7111).
10. Looper CDR Meeting Minutes, 18 November 1968 (Perkin-Elmer DAN-68-1642).

Film Path Assemblies

1. Dr. Robert E. Hufnagel interview, 28 October 1982.
2. Dr. Robert E. Hufnagel Engineering Notebook No. 1106.
3. Air Bar Vacuum Test, Breadboard Test Report No. 10
4. Air Bar Vacuum Tests, 24 February 1965 (Perkin-Elmer DAN-67-1366).
5. Gas Consumption Forecast for F' and M' Systems, 29 July 1965 (Perkin-Elmer DAN-66-0038).
6. Sensor Subsystem Monthly Technical Report No. 3, February 1967, pp. 3-4 to 3-7 (Perkin-Elmer DAN-67-0712).
7. Experimental Work Plan for Air Bars, 21 July 1967 (Perkin-Elmer DAN-67-1438).
8. Gas Flow Measurements Performed on Film Supporting Air Bars in Atmosphere and Vacuum, Technical Report TR-67-448, October 1967 (Perkin-Elmer DAN-67-1849).
9. Air Bar Design Philosophy, 29 November 1967 (Perkin-Elmer DAN-67-1973).
10. Gas Pressure Requirements for Supporting Film on Air Bars, TR-67-476, December 1967 (Perkin-Elmer DAN-68-0172).
11. Sensor Subsystem Monthly Technical Report No. 14, March 1967, p. 4-5 (Perkin-Elmer DAN-68-0436).
12. Sensor Subsystem Monthly Technical Report No. 15, April 1967, p. 4-7 (Perkin-Elmer DAN-68-0510).
13. Sensor Subsystem Monthly Technical Report No. 16, May 1968, p. 4-5 (Perkin-Elmer DAN-68-0637).
14. Sensor Subsystem Monthly Technical Report No. 18, July 1968, p. 3-3 (Perkin-Elmer DAN-68-0895).
15. Sensor Subsystem Monthly Technical Report No. 19, August 1968, p. 3-4 (Perkin-Elmer DAN-68-1403).
16. Memorandum, L.C. Smith to W.E. Brindley, 23 August 1968 (Perkin-Elmer DANX-68-7370).
17. Sensor Subsystem Monthly Technical Report No. 21, October 1968, p. 4-22 (Perkin-Elmer DAN-68-1473).
18. Development Test Plan for Air Bar Assembly, 17 January 1969 (DAN-69-0102).
19. Flux Plotting of Air Bar Flow Patterns, 21 March 1969 (Perkin-Elmer DAN-69-0443).
20. Final Report, Feasibility Study to Develop Mathematical Model and Computer Program for an Air Bar Design, May 1969 (Perkin-Elmer DANX-69-5551).

Film Path Assemblies (Continued)

21. Static Friction Test, 24 June 1969 (Perkin-Elmer DAN-69-1034).
22. Teflon Test, 8 August 1969 (Perkin-Elmer DAN-69-1002).
23. Dynamic Film - Air Bar Test Report, TR-69-620, 29 September 1969 (Perkin-Elmer DAN-69-1121).
24. Ad Hoc Study Report, Volume I, Part 4, Supplementary Report No. 115, Film Transport Mechanism and Shuttle Mechanics, 27 January 1965 (Perkin-Elmer AH65-0504).
25. Project Memorandum No. 120, Film Rollers, 7 September 1966 (Perkin-Elmer DAN-67-0611).
26. Project Memorandum No. 190, Description of Proposed Standard Roller, 4 January 1967 (Perkin-Elmer DAN-67-0163).
27. Sensor Subsystem Monthly Technical Report No. 3, February 1967, p. 3-2 (Perkin-Elmer DAN-67-0712).
28. Sensor Subsystem Monthly Technical Report No. 4, April 1967, p. 3-5 (Perkin-Elmer DAN-67-0853).
29. Sensor Subsystem Monthly Technical Report No. 7, July 1967, p. 4-2 (Perkin-Elmer DAN-67-1047).
30. Project Memorandum No. 593-X, Experimental Work Plan for Film Path Rollers, 1 November 1967 (DAN-67-1871).
31. Sensor Subsystem Monthly Technical Report No. 10, October 1967, p. 4-5 (Perkin-Elmer 67-1795).
32. Sensor Subsystem Monthly Technical Report No. 13, January 1968, p. 4-6 (Perkin-Elmer DAN-68-0213).
33. Sensor Subsystem Monthly Technical Report No. 16, May 1968, p. 4-6 (Perkin-Elmer DAN-68-0637).
34. Sensor Subsystem Monthly Technical Report No. 17, June 1968, p. 4-10 (Perkin-Elmer DAN-68-0766).
35. Sensor Subsystem Monthly Technical Report No. 18, July 1968, p. 4-8 (Perkin-Elmer DAN-68-0895).
36. Sensor Subsystem Monthly Technical Report No. 26, March 1969, p. 4-16 (Perkin-Elmer DAN-69-0534).
37. Charles D. Cowles interview, 29 September 1981.
38. Ad Hoc Report, Volume 1, Part 4, Supplementary Report No. 115, Film Transport Mechanism and Shuttle Mechanics, p. 5, 20 January 1965 (Perkin-Elmer AH65-0504).
39. Perkin-Elmer Technical Presentation to Customer, May 1965 (Perkin-Elmer AH65-0916).
40. Perkin-Elmer Presentation to Associate Contractors, 30 September 1965 (Perkin-Elmer AH65-1087).
41. Payload Contractor's Design Review Package, Revision A, 12 November 1965, p. 2-6 (Perkin-Elmer AH65-1190).
42. Payload Contractor's Presentation, 9 December 1965 (Perkin-Elmer AH65-1275).
43. Project Memorandum No. 86, Film Guidance Experiments, 13 May 1966 (PerkinElmer DAN-67-0661).
44. Project Memorandum No. 92, Crowned Rollers, 9 June 1966 (Perkin-Elmer DAN-66-0112).
45. Project Memorandum No. 93, Film Steering Cylindrical Rollers, 16 June 1966 (Perkin-Elmer DAN-66-0113).
46. Sensor Subsystem Proposal, 21 July 1966, p. 5-53 (Perkin-Elmer AH66-1401).
47. Investigation of Active Steerer Systems, 2 November 1966 (Perkin-Elmer DAN-67-0316).
48. Project Memorandum No. 140, Film Steering Devices, 7 November 1966 (Perkin-Elmer DAN-67-0610).

Film Path Assemblies (Continued)

49. Sensor Subsystem Monthly Technical Report No. 1, December 1966, p. 3-16 (Perkin-Elmer DAN-67-0160).
50. Project Memorandum No. 215, Film Guidance, 6 February 1967 (Perkin-Elmer DAN-67-0375).
51. Investigation of Passive Film Steering Using Crowned Rollers, 23 May 1967 (Perkin-Elmer DANX-67-6400).
52. Sensor Subsystem Monthly Technical Report No. 4, April 1967, p. 3-3 (Perkin-Elmer DAN-67-0853).
53. Sensor Subsystem Monthly Technical Report No. 8, August 1967, p. 4-4 (Perkin-Elmer DAN-67-1577).
54. Sensor Subsystem Monthly Technical Report No. 12, December 1967, p. 4-1 (Perkin-Elmer DAN-68-0032).
55. Project Memorandum No. 675-X, Film Low-Frequency Sideward Dynamics at Self-Aligning Air Bars, 8 January 1968 (Perkin-Elmer DAN-68-0067).
56. Film Path PDR Meeting Minutes No. 211X, 20 December 1967 (Perkin-Elmer DAN-68-0378).
57. Sensor Subsystem Monthly Technical Report No. 15, April 1968, p. 4-6 (Perkin-Elmer DAN-68-0510).
58. Project Memorandum No. 851, Study of Proposed Steerers, 21 May 1968 (Perkin-Elmer DAN-68-0594).
59. Film Path Concept Review Meeting Minutes, 19 April 1967 (Perkin-Elmer DAN-67-9834).
60. Sensor Subsystem Monthly Technical Report No. 5, May 1967, p. 4-1 (Perkin-Elmer DAN-67-1047).
61. Sensor Subsystem Monthly Technical Report No. 10, October 1967, p. 4-1 (Perkin-Elmer DAN-67-1795).
62. Sensor Subsystem Monthly Technical Report No. 12, December 1967, p. 4-1 (Perkin-Elmer DAN-68-0032).
63. Sensor Subsystem Monthly Technical Report No. 13, January 1968, p. 4-1 (Perkin-Elmer DAN-68-0213).
64. Sensor Subsystem Monthly Technical Report No. 15, April 1968, p. 4-1 (Perkin-Elmer DAN-68-0510).
65. Film Path CDR Meeting Minutes, 19 February 1969 (Perkin-Elmer DAN-69-0385).
66. Protem Proposal, 14 May 1965 (Perkin-Elmer AH65-0817).
67. Payload Section Presentation, 30 September 1965 (Perkin-Elmer AH65-1087).
68. Pneumatic Subsystem Schematic, General Electric SK4830M0048, 10 August 1965 (Perkin-Elmer AH65-1127).
69. Pressure Vessel Sizing, Technical Report 65-177, 20 August 1965 (Perkin-Elmer AH65-1026).
70. Sensor Subsystem Proposal, 21 July 1966, TR-66-300 (Perkin-Elmer AH66-1401).
71. Pneumatic Subsystem Concept Review, Project Memorandum No. 304, 21 April 1967 (Perkin-Elmer DAN-67-0804).
72. Pneumatic Subsystem Concept Review, Project Memorandum No. 304A, 4 May 1967 (Perkin-Elmer DAN 67-0804A).
73. Sensor Subsystem Monthly Technical Report No. 6, June 1967, p. 4-3 (Perkin-Elmer DAN-67-1233).
74. Sensor Subsystem Monthly Technical Report No. 7, July 1967, p. 4-5 (Perkin-Elmer DAN-67-1447).
75. Sensor Subsystem Monthly Technical Report No. 8, August 1967, p. 4-11 (Perkin-Elmer DAN-67-1577).
76. Sensor Subsystem Monthly Technical Report No. 11, November 1967, p. 4-11 (Perkin-Elmer DAN-67-2022).

Hx TOP SECRET

Film Path Assemblies (Continued)

77. Sensor Subsystem Monthly Technical Report No. 13, January 1968, p. 4-6 (Perkin-Elmer DAN-68-0213).
78. Sensor Subsystem Monthly Technical Report No. 14, March 1968, p. 4-18 (Perkin-Elmer DAN-68-0436).
79. Sensor Subsystem Monthly Technical Report No. 15, April 1968, p. 4-11 (Perkin-Elmer DAN-68-0510).
80. Sensor Subsystem Monthly Technical Report No. 16, May 1968, p. 4-8 (Perkin-Elmer DAN-68-0637).
81. Sensor Subsystem Monthly Technical Report No. 17, January 1968, p. 4-12 (Perkin-Elmer DAN-68-0766).
82. Sensor Subsystem Monthly Technical Report No. 19, August 1968, p. 4-4 (Perkin-Elmer DAN-68-1403).
83. Sensor Subsystem Monthly Technical Report No. 23, December 1968, p. 4-4 (Perkin-Elmer DAN-69-0103).
84. Ibid., p. 2-5.

Take-Up Assembly

1. Conceptual Design Approach for Take-Up Spool Assembly, Technical Report No. 66-292, 20 April 1966 (Perkin-Elmer DAN-67-0418).
2. Preliminary Design Report of the Take-Up Spools, Technical Report No. 66-333, 8 August 1966 (Perkin-Elmer DAN-66-0008).
3. Sensor Subsystem Monthly Technical Report No. 1, December 1966, p. 3-39 (Perkin-Elmer DAN-67-0160).
4. Take-Up Concept Review Meeting Minutes, 7 January 1967 (DAN-67-0264).
5. Sensor Subsystem Monthly Technical Report No. 4, April 1967, p. 3-27 (Perkin-Elmer DAN-67-0853).
6. Ibid., p. 2-25.
7. Sensor Subsystem Monthly Technical Report No. 5, May 1967, p. 4-24 (Perkin-Elmer DAN-67-1047).
8. Sensor Subsystem Monthly Technical Report No. 8, August 1967, p. 4-43 (Perkin-Elmer DAN-67-1577).
9. PM No. 547-X, Review of RCA Proposal for Take-Up Subsystem, 5 September 1967 (Perkin-Elmer DAN-67-1575).
10. Take-Up Subsystem Concept Review Meeting Minutes, 15 November 1967 (Perkin-Elmer DAN-67-2087).
11. Sensor Subsystem Monthly Technical Report No. 11, November 1967, p. 2-5 (Perkin-Elmer DAN-67-2022).
12. Take-Up Preliminary Design Review Meeting Minutes, 20 February 1968 (Perkin-Elmer DAN-68-0356).
13. Sensor Subsystem Monthly Technical Report No. 20, September 1968, p. 2-9 (DAN-68-1246).

System Electronics

1. Protem Proposal, 14 May 1965 (Perkin-Elmer AH65-0817).
2. Film Transport Functional Description, Technical Report No. 65-72, 5 May 1965 (Perkin-Elmer AH65-0780).
3. On-Board Diagnostic Sensing, Technical Report No. 65-87, 12 May 1965 (Perkin-Elmer AH65-0812).

Hx TOP SECRET

System Electronics (Continued)

4. System Control and Synchronization, Technical Report No. 66-291, 25 April 1966 (Perkin-Elmer DAN-66-0042).
5. Protem Presentation (Perkin-Elmer AH65-0916).
6. Payload Contractor's Presentation, 9 December 1965 (Perkin-Elmer AH65-1275).
7. Perkin-Elmer Design Definition Sensor Subsystem, Volume 1 (Part 1), 21 July 1966 (Perkin-Elmer AH66-1401).
8. Richard H. Carricato interview, 22 September 1982.

Systems Engineering

1. Payload Contractor's Presentation, 9 December 1965, Program Organization (Perkin-Elmer AH65-1275).
2. Sensor Subsystem Negotiation Handbook, Organization Chart, p. I-4, 6 March 1968 (Perkin-Elmer DAN-68-0315).
3. Sensor Subsystem PDR Meeting Minutes, 29 February-1 March 1968 (Perkin-Elmer DAN-68-0515).
4. Memorandum, Donald W. Patterson to Kenneth Patrick, 6 May 1968 (Perkin-Elmer DANX-68-6626).
5. System CDR Presentation, 5 & 6 March 1969 (Perkin-Elmer DAN-69-0367).
6. Design Audit Technical Report, Seven Priority Design Risk Area, 19 June 1968 (Perkin-Elmer DAN-68-0740).
7. Thermal Control and Pressure Requirements, Technical Memorandum No. 2, 20 November 1964 (Perkin-Elmer AH65-0575).
8. Interim Test Report-Film In Vacuum, Technical Report No. 66-245 (Perkin-Elmer AH65-1286).
9. Film In Vacuum, Final Test Report, Technical Report No. 66-294, 25 April 1966 (Perkin-Elmer AH66-1377).
10. Floating Film, Technical Report No. 66-242, 28 February 1966 (Perkin-Elmer AH66-1283).
11. Sensor Subsystem Design Definition, Volume 1 (Part 1), 21 July 1966, p. D-1 (Perkin-Elmer AH66-1401).
12. System Pressurization, Project Memorandum No. 130, 30 September 1966 (Perkin-Elmer DAN-67-0312).
13. System Pressurization, Project Memorandum No. 164, 3 December 1966 (Perkin-Elmer DAN-66-0088).
14. Sensor Subsystem Monthly Technical Report No. 1, December 1966, p. 3-4 (Perkin-Elmer DAN-67-0160).
15. Program Plan - Design Studies of a Pressurized Film Transport Subsystem, 28 December 1966 (Perkin-Elmer DANX-67-6023).
16. Film Moisture Content - Recommended Test to Determine Variations, 29 December 1966 (Perkin-Elmer DANX-67-6019).
17. Pressurization of Sensor Subsystem - Initial Evaluation of Problems, 4 January 1967 (Perkin-Elmer DANX-67-6017).
18. Film Path Pressurization Meeting Minutes, 25 January 1967 (Perkin-Elmer DAN-67-0308).
19. Interaction of the Film and Its Environment, Technical Report TR67-373, 24 February 1967 (Perkin-Elmer DAN-67-0535).
20. Sensor Subsystem Monthly Technical Report No. 3, February 1967, p. 2-15 (Perkin-Elmer DAN-67-0712).
21. Sensor Subsystem Monthly Technical Report No. 4, April 1967, p. 2-14 (Perkin-Elmer DAN-67-0853).

Systems Engineering (Continued)

22. Sensor Subsystem Monthly Technical Report No. 4, p. 3-2.
23. Preliminary Experimental Work Plan for Abbreviated Film Path, Project Memorandum No. 289, 4 April 1967 (Perkin-Elmer DAN-67-0706).
24. Film Path Pressurization, Design Approach, Project Memorandum No. 295, 11 April 1967 (Perkin-Elmer DAN-67-0738).
25. Technical Discussion Regarding Pressurization Meeting Minutes No. 108, 31 May 1967 (Perkin-Elmer DAN 67-1164).
26. Sensor Subsystem Monthly Technical Report No. 7, July 1967, p. 4-4 (Perkin-Elmer DAN-67-1447).
27. Sensor Subsystem Monthly Technical Report No. 16, May 1968, p. 4-74 (Perkin-Elmer DAN-68-0637).
28. Project Memorandum, L. C. Smith to W. E. Brindley, 16 May 1968 (Perkin-Elmer DANX-68-6704).
29. Project Memorandum, L. C. Smith to W. E. Brindley, 11 June 1968 (Perkin-Elmer DANX-68-6840).
30. Sensor Subsystem Monthly Technical Report No. 17, June 1968, p. 2-25 (Perkin-Elmer DAN-68-0766).
31. Sensor Subsystem Monthly Technical Report No. 20, September 1968, p. 1-1 (Perkin-Elmer DAN-68-1246).
32. Perkin-Elmer Design Definition Sensor Subsystem, Volume 1 (Part 1), 21 July 1966 (Perkin-Elmer AH66-1401).
33. Ibid., p. 5-55.
34. Ibid., p. 12-2.
35. Sensor Subsystem Monthly Technical Report No. 1, December 1966, p. 3-19 (Perkin-Elmer DAN-67-0160).
36. Film Tracking Analysis, Technical Report No. 67-477, 17 November 1967 (Perkin-Elmer DAN-67-2120).
37. Analysis of System Film Tracking Error, Technical Report No. 67-442A, 5 December 1967 (Perkin-Elmer DAN-67-2008A).
38. System Film Tracking Errors, Technical Report No. 69-581, 14 February 1969 (Perkin-Elmer DAN-69-0274).
39. Sensor Subsystem Monthly Technical Report, November 1969, p. 1-1 (Perkin-Elmer DAN-69-1323).
40. Ibid., p.1-3.
41. Sensor Subsystem Monthly Technical Report No. 42, July 1970, p. 2-2 (BIF 007-0545-70).
42. Ibid., p. 2-1.
43. Sensor Subsystem Monthly Technical Report No. 43, August 1970 (BIF 007-0582-70).
44. Sensor Subsystem Monthly Technical Report No. 44, September 1970 (BIF 007-0621-70).
45. Coarse Path Tracking in the Hexagon Camera System, PFS Technical Report No. 1, January 1973 (BIF 007-3298-73).
46. Memorandum, D. W. Patterson to Michael F. Maquire, 24 September 1970 (BIF 007-6001-70).
47. Sensor Subsystem Monthly Technical Report No. 50, March 1971, p. 5-35 (BIF 007-0522-71).
48. Sensor Subsystem Monthly Technical Report No. 51, April 1970, p. 2-1 (BIF 007-0673-71).
49. Sensor Subsystem Monthly Technical Report No. 52, May 1971, p. 5-1 (BIF 007-0878-71).
50. Ibid., pp. 7-30 and 7-34.

Systems Engineering (Continued)

51. Sensor Subsystem Monthly Technical Report No. 53, June 1971, p. 2-1 (BIF 007-1063-71).
52. Ibid., p. 5-2.
53. Sensor Subsystem Monthly Technical Report No. 57, October 1971, p. 7-2 (BIF 007-1671-71).
54. System Performance Evaluation Team - Mission 1203, 6 November 1972, p. 4-1 (BIF 007-3297-73).
55. Ibid., p. 6-1.
56. Ibid., p. 2-2.
57. Memorandum, D. W. Patterson to L. M. Weeks, 24 May 1973 (BIF 007-4308-73).
58. Memorandum, L. M. Weeks to D. W. Patterson, 6 June 1973 (BIF 007-0642-73).
59. SV-11 (S/N 014) Tracking Anomaly Investigations, Project Memorandum 1553X, 1 October 1975 (BIF 007-0576-75).
60. Arnold Wallace interview, 18 August 1983.
61. Film Properties Meeting Minutes, 6 December 1966 (Perkin-Elmer DAN-66-0109).
62. Sensor Subsystem Monthly Technical Report No. 21, October 1968, p. 2-20 (Perkin-Elmer DAN-68-1473).
63. Sensor Subsystem Monthly Technical Report No. 27, April 1969, p. 2-12 (Perkin-Elmer DAN-69-0694).
64. Sensor Subsystem Monthly Technical Report No. 32, September 1969, p. 4-1 (Perkin-Elmer DAN-69-1163).
65. Sensor Subsystem Monthly Technical Report No. 35, December 1969, p. 4-2 (Perkin-Elmer DAN-70-0033).
66. Roger J.P. Gaulin interview, 29 October 1982.
67. Project Memorandum No. 211, Justification for 6 x 8 Foot Chamber for Film Path Environment Experiments, 20 January 1967 (Perkin-Elmer DAN-67-0277).
68. Sensor Subsystem monthly Technical Report No. 3, February 1967, p. 2-19 (Perkin-Elmer DAN-67-0783).
69. Project Memorandum No. 289, Preliminary Experimental Work Plan for Abbreviated Film Path, 4 April 1967 (Perkin-Elmer DAN-67-0706).
70. Sensor Subsystem Monthly Technical Report No. 7, July 1967, p. 3-4 (Perkin-Elmer DAN-67-1447).
71. Sensor Subsystem Monthly Technical Report No. 13, January 1968, p. 3-1 (Perkin-Elmer DAN-68-0213).
72. Sensor Subsystem Monthly Technical Report No. 17, June 1968, p. 3-2 (Perkin-Elmer DAN-68-0766).
73. Sensor Subsystem Monthly Technical Report No. 21, October 1968, p. 3-1 (Perkin-Elmer DAN-68-1473).
74. Sensor Subsystem Monthly Technical Report No. 22, November 1968, p. 3-4 (Perkin-Elmer DAN 68-1650).
75. Sensor Subsystem Monthly Technical Report No. 26, February 1969, p. 3-8 (Perkin-Elmer DAN-69-0534d).
76. Edward C. Mathews interview, 19 November 1982.

System Reliability

1. Sensor Subsystem Design Definition, Volume 1 (Part 1), 21 July 1966, p. 6-2 (Perkin-Elmer AH66-1401).
2. Sensor Subsystem Monthly Technical Report No. 2, January 1967, p. 6-1 (Perkin-Elmer DAN-67-0372).
3. Stanley T. Karachuk interview, 8 November 1982.

System Reliability (Continued)

4. Sensor Subsystem Monthly Technical Report No. 11, November 1967, p. 6-2 (Perkin-Elmer DAN-67-2022).
5. Sensor Subsystem Monthly Techncial Report No. 13, January 1968, p. 6-2 (Perkin-Elmer DAN-68-0213).
6. Sensor Subsystem Monthly Techncial Report No. 23, December 1968, p. 6-1 (Perkin-Elmer DAN-69-0103).
7. Letter, Major General John E. Kulpa, Jr. to Robert H. Sorensen, 14 July 1978 (BIF 007-4133-78).
8. Review of Hexagon Camera System Reliability (BIF 007-4399-78).

Manufacturing and Tests

1. Sensor Subsystem Design Definition, Volume 1 (Part 1), 21 July 1966, p. 9-16 (Perkin-Elmer AH66-1401).
2. Sensor Subsystem Monthly Technical Report No. 3, February 1967, p. 2-1 (Perkin-Elmer DAN-67-0783).
3. Sensor Subsystem Monthly Technical Report No. 9, September 1967, p. 8-2 (Perkin-Elmer DAN-67-1691).
4. Sensor Subsystem Monthly Technical Report No. 12, December 1967, p. 8-1 (Perkin-Elmer DAN-68-0032).
5. Sensor Subsystem Monthly Technical Report No. 13, January 1968, p. 8-1 (Perkin-Elmer DAN-68-0213).
6. Sensor Subsystem Monthly Technical Report No. 23, December 1968, p. 8-1 (Perkin-Elmer DAN-69-0103).
7. Sensor Subsystem Monthly Technical Report No. 27, April 1969, p. 1-1 (Perkin-Elmer DAN-69-0694).
8. Sensor Subsystem Monthly Technical Report No. 38, March 1970, p. 7-5 (Perkin-Elmer DAN-60-0281).
9. Sensor Subsystem Monthly Technical Report No. 13, January 1968, p. 8-1 (Perkin-Elmer DAN-68-0213).
10. Sensor Subsystem Monthly Technical Report No. 14, March 1968, p. 8-1 (Perkin-Elmer DAN-68-0436).
11. Sensor Subsystem Monthly Technical Report No. 15, April 1968, p. 8-1 (Perkin-Elmer DAN-68-0510).
12. Sensor Subsystem Monthly Technical Report No. 28, May 1969, p. 1-1 (Perkin-Elmer DAN-69-0800).
13. Sensor Subsystem Monthly Technical Report No. 29, June 1969, p. 1-1 (Perkin-Elmer DAN-69-0908).
14. Sensor Subsystem Monthly Technical Report No. 30, July 1969, p. 1-1 (Perkin-Elmer DAN-69-1005).
15. Sensor Subsystem Monthly Technical Report No. 31, August 1969, p. 7-1 (Perkin-Elmer DAN-69-1098).
16. Sensor Subsystem Monthly Technical Report No. 32, September 1969, p. 7-1 (Perkin-Elmer DAN-69-1163).
17. Sensor Subsystem Monthly Technical Report No. 34, November 1969, p. 1-1 (Perkin-Elmer DAN-69-1323).
18. Sensor Subsystem Monthly Technical Report No. 35, December 1969, p. 1-2 (Perkin-Elmer DAN-70-0033).
19. Sensor Subsystem Monthly Technical Report No. 36, January 1970, p. 1-1 (Perkin-Elmer DAN-70-0118).

Manufacturing and Tests (Continued)

20. Sensor Subsystem Monthly Technical Report No. 37, February 1970, p. 2-1 (Perkin-Elmer DAN-70-0194).
21. Sensor Subsystem Monthly Technical Report No. 38, March 1970, p. 3-1 (Perkin-Elmer DAN-70-0281).
22. Sensor Subsystem Monthly Technical Report No. 39, April 1970, p. 2-1 (Perkin-Elmer BIF 007-0354-70).
23. Sensor Subsystem Monthly Technical Report No. 32, September 1969, p. 1-2 (Perkin-Elmer DAN-69-1163).
24. Sensor Subsystem Monthly Technical Report No. 33, October 1969, p. 1-1 (Perkin-Elmer DAN-69-1251).
25. Sensor Subsystem Monthly Technical Report No. 34, November 1969, p. 1-1 (Perkin-Elmer DAN-69-1323).
26. Sensor Subsystem Monthly Technical Report No. 35, December 1969, p. 8-20 (Perkin-Elmer DAN-70-0033).
27. Sensor Subsystem Monthly Technical Report No. 36, January 1970, p. 1-1 (Perkin-Elmer DAN-70-0118).
28. Sensor Subsystem Monthly Technical Report No. 37, February 1970, p. 1-1 (Perkin-Elmer DAN-70-0194).
29. Sensor Subsystem Monthly Technical Report No. 39, April 1970, p. 3-1 (Perkin-Elmer BIF 007-0354-70).
30. Sensor Subsystem Monthly Technical Report No. 43, August 1970, p. 6-2 (Perkin-Elmer BIF 007-0528-70).
31. Sensor Subsystem Monthly Technical Report No. 44, September 1970, p. 6-2 (Perkin-Elmer BIF 007-0621-70).
32. Sensor Subsystem Monthly Technical Report No. 45, October 1970, p. 6-4 (Perkin-Elmer BIF 007-0696-70).
33. Sensor Subsystem Monthly Technical Report No. 46, November 1970, p. 6-6 (Perkin-Elmer BIF 007-0698-70).
34. Sensor Subsystem Monthly Technical Report No. 47, December 1970, p. 6-7 (Perkin-Elmer BIF 007-0026-71).
35. Sensor Subsystem Monthly Technical Report No. 48, January 1971, p. 6-8 (Perkin-Elmer BIF 007-0125-71).
36. Sensor Subsystem Monthly Technical Report No. 50, March 1971, p. 6-4 (Perkin-Elmer BIF 007-0522-71).
37. Sensor Subsystem Monthly Technical Report No. 37, April 1971, p. 6-4 (Perkin-Elmer BIF 007-0673-71).
38. Sensor Subsystem Monthly Technical Report No. 38, May 1971, p. 8-4 (Perkin-Elmer BIF 007-0878-71).
39. Sensor Subsystem Monthly Technical Report No. 39, June 1971, p. 9-1 (Perkin-Elmer BIF 007-1063-71).
40. Sensor Subsystem Monthly Technical Report No. 24, January 1969, p. 1-4 (Perkin-Elmer DAN-69-0322).
41. Sensor Subsystem Monthly Technical Report No. 36, January 1970, p. 1-3 and p. 8-20 (Perkin-Elmer DAN-70-0118).
42. Sensor Subsystem Monthly Technical Report No. 38, March 1970, p. 4-1 (Perkin-Elmer DAN-70-0281).
43. Sensor Subsystem Monthly Technical Report No. 39, April 1970, p. 4-1 (Perkin-Elmer BIF 007-0354-70).
44. Sensor Subsystem Monthly Technical Report No. 40, May 1970, p. 4-1 (Perkin-Elmer BIF 007-0426-70).
45. Sensor Subsystem Monthly Technical Report No. 41, June 1970, p. 4-1 (Perkin-Elmer BIF 007-0500-70).

Hx TOP SECRET

Manufacturing and Tests (Continued)

46. Sensor Subsystem Monthly Technical Report No. 42, July 1970, p. 2-1 (Perkin-Elmer BIF 007-0545-70).
47. Sensor Subsystem Monthly Technical Report No. 43, August 1970, p. 2-1 (Perkin-Elmer BIF 007-0582-70).
48. Sensor Subsystem Monthly Technical Report No. 44, September 1970, p. 2-1 (Perkin-Elmer BIF 007-0621-70).
49. Sensor Subsystem Monthly Technical Report No. 45, October 1970, p. 2-1 (Perkin-Elmer BIF 007-0697-70).

Optical Fabrication

1. Protem Proposal, 14 May 1965 (Perkin-Elmer AH65-0817).
2. Technical Memorandum, Lightweight Mirror Configurations, 9 November 1964 (Perkin-Elmer AH65-0265).
3. Principles of Optics, Presented at Wright-Patterson Air Force Base, Dayton, Ohio, 10-11 January 1962.
4. Contract Negotiation Handbook, 6 March 1968, pg. I-14 (Perkin-Elmer DAN-68-0315).
5. Sensor Subsystem Design Definition, Volume 1 (Part 1), 21 July 1966, p. 9-6 (Perkin-Elmer AH66-1401).
6. TWX Message ███████████, Biweekly Report, 25 October 1966.

SECTION 4 RELATIONSHIPS AND INTERFACES WITH ASSOCIATE CONTRACTORS AND SUBCONTRACTORS

Associate Contractors and Responsibilities

1. Memorandum, Leslie C. Dirks to Earle Brown, Spacecraft Preliminary Design and Project Program Schedule (Attachment #1 to original RFP), 8 July 1964 (Perkin-Elmer DANX-67-6167).

Selection of Subcontractors

1. Technical Report No. 65-81, Identification of Vendors Suited to F-Prime Contract Requirements, 10 May 1965 (Perkin-Elmer AH65-0789).
2. Program Plan Sensor Subsystem, Volume II, p. M-1, 21 July 1966 (Perkin-Elmer AH66-1402).
3. Ralph E. Sisk interview, 27 January 1981.
4. Security Evaluation of Subcontractors to Perkin-Elmer, Optical Technology Division, 29 September 1967 (Perkin-Elmer DAN-67-1626).
5. Status of Major Subcontracts, 13 June 1968 (Perkin-Elmer DAN-68-0695).
6. Sisk interview.

Development of Interfaces

1. Photographic System Specification, Part 3, 26 January 1965 (Perkin-Elmer AH65-0503).
2. System Specification Book, 16 November 1965 (Perkin-Elmer AH65-1211).
3. TWX Message ████████, 10 March 1965.
4. Memorandum, Summary of Interface Meetings, 29 July 1965 (Perkin-Elmer AH65-1030-1).
5. Memorandum, Electrical Interface Meeting, 29 September 1965 (Perkin-Elmer AH65-1030-5).
6. Payload Contractor's Presentation, 9 December 1965 (Perkin-Elmer AH65-1275).
7. Request for Proposal for Sensor Subsystem, General Search and Surveillance System, p. 6, 19 May 1966 (Perkin-Elmer AH66-1400).
8. Interface and Liaison Group Program Plan, 6 October 1966 (Perkin-Elmer AH66-1484).
9. Interface Control Detail Manual, 14 October 1966 (Perkin-Elmer AH65-1485).
10. Interface Management Manual and Interface Control Procedures, August 1968 (Perkin-Elmer DANX-69-4054).
11. Design Definition Sensor Subsystem, Volume 1 (Part 1), p. 8-1 (Perkin-Elmer AH66-1401).
12. Memorandum, Preliminary Interface Meetings, 26 October 1966 (Perkin-Elmer AH66-1609).
13. Meeting Memorandum No. 138, SS/SBA Interface, 31 August 1967 (Perkin-Elmer DAN-67-1595).
14. AVE Interface Requirements for Sensor Subsystem, IRD 501, 31 March 1967 (Perkin-Elmer DANX-67-6215).
15. System CDR Presentation, 6 March 1969 (Perkin-Elmer DAN-69-0367).
16. Memorandum, Completion of Phaseover of Interface Responsibility from SETS to SSC, 9 July 1968 (Perkin-Elmer DANX-68-6948).
17. George R. Gray interview, 18 January 1983.

SECTION 5 SYSTEM INTEGRATION, LAUNCH, ORBITAL OPERATIONS AND RECOVERY

Development of the West Coast Field Office

1. Lincoln Endelman interview, 3 December 1981.
2. Charles O. Bryant interview, 18 October 1982.
3. Endelman interview.
4. Memorandum for the record, Christopher Fitzgerald, Review of West Coast Activities, 26 July 1968 (Perkin-Elmer DANX-68-7195).
5. Memorandum, H.J. Loper to C.O. Bryant, 25 July 1971, BIFX 007-3291-71.
6. Sensor Subsystem Monthly Technical Report No. 25, February 1969, p. 10-2 (Perkin-Elmer DAN-69-0432).
7. Sensor Subsystem Monthly Technical Report No. 34, November 1969, p. 13-5 (Perkin-Elmer DAN-69-1323).
8. Sensor Subsystem Monthly Technical Report No. 26, March 1969, p. 13-2 (Perkin-Elmer DAN-69-0534).
9. Sensor Subsystem Monthly Technical Report No. 36, January 1970, p. 13-1 (Perkin-Elmer DAN-60-0118).

Final Assembly and Testing of Flight Model 1 (SV-1)

1. Project Memorandum, PM-976-X, Midsection Transportation Between SSC and SAFB, 19 September 1968 (Perkin-Elmer DAN-68-1127).
2. Sensor Subsystem Monthly Technical Report No. 35, December 1969, p. 13-1 (Perkin-Elmer DAN-70-0033).
3. Sensor Subsystem Monthly Technical Report No. 45, October 1970, p. 2-2, BIF 007-0697-70.
4. Sensor Subsystem Monthly Technical Report No. 46, November 1970, p. 2-1, BIF 007-0798-70.
5. Ibid, p. 6-5.
6. Sensor Subsystem Monthly Technical Report No. 47, December 1970, p. 2-1, BIF 007-0026-71.
7. Sensor Subsystem Monthly Technical Report No. 48, January 1971, p. 2-1, BIF 007-0125-71.
8. Sensor Subsystem Monthly Technical Report No. 49, February 1971, p. 2-1, BIF 007-0322-71.
9. Sensor Subsystem Monthly Technical Report No. 50, March 1971, p. 2-1, BIF 007-0522-71.
10. Sensor Subsystem Monthly Technical Report No. 51, April 1971, p. 2-1, BIF 007-0673-71.
11. Sensor Subsystem Monthly Technical Report No. 52, May 1971, p. 2-1, BIF 007-0878-71.
12. Shipping Certification, SV-1, 4 June 1971, BIFX 007-4813-71.
13. Sensor Subsystem Monthly Technical Report No. 53, June 1971, p. 2-1, BIF 007-1063-71.
14. Launch Certification for SV-1, 14 June 1971, BIFX 007-5402-71.

Mission Activities

1. Sensor Subsystem Monthly Technical Report No. 53, June 1971, p. 2-1 and p. 9-3, BIF 007-1063-71.
2. Sensor Subsystem Monthly Technical Report No. 54, July 1971, p. 2-1, BIF 007-1175-71.
3. SV-1 Sensor System Flight Anomalies, 4 December 1971, BIFX 007-6631-71.

Recovery of the Lost RV-3

1. Sensor Subsystem Monthly Technical Report No. 54, July 1971, p. 2-2 (BIF 007-1175-71).
2. Memorandum, D.W. Patterson for record, 28 July 1971 (BIFX 007-5333-71).
3. Memorandum ME-45, Leonard B. Molaskey, 2 August 1971.
4. Memorandum ME-50, Leonard B. Molaskey to Project File, 5 August 1971 (BIF 007-1172-71).
5. Memorandum ME-49, Leonard B. Molaskey to Project File, 5 August 1971 (BIF 007-1171-71).
6. Sensor Subsystem Monthly Technical Report No. 55, August 1971, p. 2-7 (BIF 007-1367-71).
7. Patterson Memorandum.
8. Memorandum ME-57, Leonard B. Molaskey to Project File, 15 September 1971.
9. Memorandum ME-59, Leonard B. Molaskey to Project File, 24 September 1971.
10. Sensor Subsystem Monthly Technical Report No. 56, September 1971, p. 2-2 (BIF 007-1522-71).
11. Memorandum ME-60, Leonard B. Molaskey to Project File, 30 September 1971.
12. Memorandum ME-61, Leonard B. Molaskey to Project File, 29 September 1971.
13. Memorandum ME-62, Leonard B. Molaskey to Project File, 4 October 1971.
14. Memorandum ME-64, Leonard B. Molaskey to Project File, 8 October 1971.
15. Memorandum ME-65, Leonard B. Molaskey to Project File, 12 October 1971.
16. Memorandum ME-66, Leonard B. Molaskey to Project File, 13 October 1971.
17. Memorandum ME-70, Leonard B. Molaskey to Project File, 18 November 1971 (BIF 007-1717-71).
18. Memorandum ME-71, Leonard B. Molaskey to Project File, 22 November 1971 (BIF 007-1743-71).
19. Memorandum ME-72, Leonard B. Molaskey to Project File, 10 December 1971 (BIF 007-1829-71).
20. Memorandum ME-80, Leonard B. Molaskey to Project File, 31 January 1972 (BIF 007-0155-72).

www.ingramcontent.com/pod-product-compliance
Lightning Source LLC
Chambersburg PA
CBHW082118230426
43671CB00015B/2730